Studying Engineering

A Road Map to a Rewarding Career

Fourth Edition

by
Raymond B. Landis, Dean Emeritus
College of Engineering, Computer Science, and Technology
California State University, Los Angeles

Published by:
Discovery Press
Los Angeles, California
www.discovery-press.com

Permissions and Copyrights

Cover design by Dave McNutt

Chapter title illustrations by Brian Jefcoat

Graphic illustrations by Kerry Lampkin

Case study of University of Maryland Gamera human-powered helicopter project by permission of Inderjit Chopra, Director of the Alfred Gessow Rotorcraft Center

Franklin Chang-Diaz photo in Chapter 1 courtesy of National Aeronautics and Space Administration

Studying Engineering: A Road Map to a Rewarding Career, Fourth Edition

Discovery Press/2013

10 9 8 7 6 5

ISBN 978-0-9793487-4-7

Inquiries and comments should be addressed to:

Raymond B. Landis, Ph.D.
Dean Emeritus of Engineering, Computer Science, and Technology
California State University, Los Angeles
Los Angeles, California 90032
E-mail: rlandis@calstatela.edu

Distributed by:

Legal Books Distributing
4247 Whiteside Street
Los Angeles, CA 90063
Website: www.legalbooksdistributing.com
Telephone: (323) 526-7110
            (800) 200-7110 (From outside Los Angeles County)
E-mail: info@legalbooksdistributing.com

Books may be ordered by e-mail, by telephone, or on-line

TO KATHY

FOREWORD

When I was a sophomore in high school, I decided that I wanted to be a chemical engineer when I grew up. I could invent all sorts of reasons for this decision that would make me sound like an unusually wise and thoughtful 15-year-old, but they would all be lies.

The truth is at the time there was a great job market for engineers, and stories of red carpets and multiple job offers and outlandishly high starting salaries were laid on us regularly by teachers and counselors – and in my case, by my parents. Just about every boy who could get a *"B"* or better in math and science courses decided that he was born to be an engineer, and I saw no reason to buck the trend. Why *chemical* engineering? Because – sadly, this is also the truth – I had gotten a chemistry set for my birthday, and I thought pouring one liquid into another and having it turn green was seriously cool.

Like most of my engineering-bound classmates, I knew nothing about what engineers actually did for a living, and when I enrolled in chemical engineering at the City College of New York two years later, I still knew nothing. There was a freshman orientation course, but it was just the old "Sleep 101" parade of unenthusiastic professors delivering dreary 40-minute sermons about civil engineering, mechanical engineering, and so forth. It's a wonder that this course didn't drive more students away from engineering than it motivated. Perhaps it did.

My ignorance persisted for pretty much the next three years as I worked through the math and physics and chemistry and thermo and transport and circuits you have to know to graduate in engineering but constitute only a small fraction of what engineers actually do. It wasn't until I got into the unit operations lab in my fourth year and then spent a summer in industry that I started to get a clue about what engineering is really about – figuring out why things aren't working the way they're supposed to and fixing them, and designing and building other things that work better or work just as well but cost less.

And what engineers did for a living was only the tip of the iceberg of what I didn't know as a freshman. In high school I rarely cracked a textbook and still came out with nearly straight *"A's"*, but it only took one college physics exam to let me know that the game had changed. I also left high school thinking I was a great writer, but the *"D+"* on my first English paper set me straight about that too. Plus, I didn't know how to take effective notes, summarize long reading assignments, prepare for and take tests, or strike a good balance between school and the rest of my life. I could go on but you get the idea.

I eventually figured it out, of course. If I hadn't, I wouldn't have graduated and gone on to become an engineering professor and the author of this foreword. Unfortunately, many of my classmates never did get it, and most of them were gone by the end of the second year. And I know they had the ability to succeed.

I don't think engineering school should be an academic obstacle course designed to weed out students who have the ability to succeed but lack basic study skills. If something is important for students to know, there's nothing wrong with giving them some guidance in figuring it out. We do that routinely with math and science and control and design. Why not do it with studying and learning?

That's where Ray Landis and *Studying Engineering* come in. The book is a compendium of everything I wish someone had told me during my freshman year of college. If I could have read it then, even if I had only absorbed a fraction of the wisdom it contains, I would have been spared the major headache of having to learn it the hard way. And if the book had been used in a first-year engineering course taught by a knowledgeable and supportive instructor, the next four years of my life would have been far less stressful, and many of my talented classmates who dropped out as freshmen and sophomores would instead have graduated with me.

Virtually everything students need to know to succeed in engineering school is detailed in *Studying Engineering*. Using a conversational tone and numerous real-world examples and anecdotes, Professor Landis paints a vivid picture of the vast range of things engineers do, the world-changing things they have done in the past, and the challenges to problem-solving ability and creativity that engineers routinely face. He also introduces students to the learning process – how it works, when and why it goes wrong, and how to avoid the pitfalls that have ensnared generations of engineering students (including those unfortunate classmates of mine).

Moreover, *Studying Engineering* introduces its readers to themselves and to one another, providing insights into different ways people approach learning tasks and respond to instruction. Students who take this material to heart will gain a better understanding of their own strengths and weaknesses, learning ways to capitalize on the former while overcoming the latter. Their new knowledge will also improve their ability to communicate with their classmates, teammates, and professors. These insights and skills will serve them well throughout college and their subsequent professional careers, whether or not they remain in engineering.

If you are an engineering educator who teaches first-year students, I invite you to think about the things you wish someone had told you when you were a freshman and then use *Studying Engineering* to help convey those messages.

If you are a student, I encourage you to pay close attention to the book because it will teach you how to get the most out of your engineering education. If you're going to succeed and excel in engineering school and down the road as an engineering professional, you'll need to learn these things sooner or later. My advice is to do it sooner.

Richard M. Felder
Hoechst Celanese Professor Emeritus of Chemical Engineering
North Carolina State University
Raleigh, North Carolina

TABLE OF CONTENTS

PREFACE

Studying Engineering has been updated and expanded. Dated material has been updated and a wealth of relevant Internet sites has been added. Substantial new graphics have been added as well to improve readability. A new Prologue has been included to give students a clearer perspective on what this book has to offer and – more importantly – what steps they can take to get the most from it. New sections have been added on subjects such as fixed vs. growth mindset, reverse engineering, sustainability, life-long learning, study abroad, entrepreneurship, and teamwork and leadership.

The Prologue, "What This Book Has to Offer and How to Get It," discusses the potential of this book to make a difference in students' lives, and provides guidance on how to realize that potential.

Chapter 1 lays the foundation for the book by introducing and overviewing the process of achieving success in engineering study. Key elements of the success process – goal identification, goal clarification, and behavioral and attitudinal change – are presented. Three models that will help students understand what is meant by a quality education and how to go about getting that education are also introduced. The chapter closes with the important topic of "Structuring Your Life Situation."

Chapter 2 addresses the subject of professional development. A primary purpose of the chapter is to motivate students through an increased understanding of the engineering profession and an awareness of the rewards and opportunities that will come to them if they are successful in their engineering studies. The University of Maryland's "Gamera" human-powered helicopter project is used to bring the engineering design process to life. The National Academy of Engineering's Grand Challenges for Engineering is used to show some of the many exciting problems engineers will need to tackle in the future.

Chapter 3 provides an overview of the *teaching/learning* process. Various types of Learning modes – cognitive, psychomotor, and affective – are described. Preferred learning styles and teaching styles are also discussed. Students are given general guidelines to strengthen their learning process and a summary of the most common mistakes students make is presented, along with ways to avoid these mistakes.

Chapter 4 provides guidance on how to get the most out of the teaching process. The chapter emphasizes the importance of getting off to a good start and discusses strategies for taking full advantage of lectures – including listening skills, note-taking skills, and questioning skills. Approaches for making effective use of professors are described in detail.

Chapter 5 guides students in designing their learning process. Two important skills for learning – reading for comprehension and analytical problem solving – are covered. Approaches for organizing the learning process, such as time management skills, are also discussed. Study skills relevant to math/science/engineering coursework are emphasized. Finally, ways to make effective use of peers through collaborative learning and group study are also described.

Chapter 6 focuses on the important subject of personal growth and development. A *Student Success Model* is presented to help students understand the process of making behavioral and attitudinal changes essential to success in engineering study. Important personal development topics – understanding self, appreciating differences, personal assessment, communication skills, and health and wellness – are included as well. Finally, a section has been added on the important topic of leadership and teamwork.

Chapter 7 addresses five extracurricular activities that can greatly enhance the quality of a student's education: (1) student organizations, (2) engineering projects, (3) pre-professional employment, (4) study abroad, and (5) service to the university.

Chapter 8 provides an orientation to the engineering education system: faculty, curriculum, students, facilities, administration, and institutional commitment. Academic regulations, student ethics, and opportunities for graduate education in engineering are also covered in this chapter. We close with a discussion of engineering as a means of preparation for further education in business, law, and medicine.

Appendices are devoted to five important topics: 1) Design Project; 2) Definitions of Engineering; 3) Greatest Engineering Achievements of the 20th Century; 4) Engineers among the World's 200 Wealthiest Individuals; and 5) Description of Engineering Disciplines.

The target audience for the book is first-year engineering students; therefore, it is ideally suited for use in an *Introduction to Engineering* course that has a "student development/student success" objective. Much of what is in the book has direct application to the community college experience, and the topics that are specific to the four-year university experience can provide community college students with a preview of what they will encounter when they transfer to a four-year institution.

High school students considering engineering as their college major will find the book useful as well. Engineering faculty can turn to it as a resource for ideas they can convey to students in formal and informal advising sessions or in the classroom. Deans of engineering have indicated that the book contains material that is helpful in preparing talks they give to high school students and first-year engineering students.

This book was the outgrowth of more than 30 years of teaching *Introduction to Engineering* courses. Much of the material was developed through brainstorming exercises with students. My greatest thanks go to the many students who contributed to the evolution of the ideas in this book. Thanks also go to the many engineering professors who have used the book since the *First Edition* was published in 1995. Those who provided valuable feedback used to improve this edition include: Dom Dal Bello, Rich Bankhead, Zahir Khan, David Gray, Jack Hopper, Sami Maalouf, Bill Latto, Nick Arnold, Zanj Avery, Ali Kujoory, Julie Zhao, Artin Davidian, Jawa Mariappan, Jeff Froyd, Anthony Donaldson, Janet Meyer, Dave Kaeli, Thalia Anagnos, Herb Schroeder, Bev Louie, Marty Wood, and Kevin McLaughlin.

Many people contributed directly or indirectly to the creation of the book – both its original and its revised form. Much credit goes to my partner Martin Roden for encouraging me to self-publish the book and for his constant help and support. Great thanks to Dave McNutt for putting his extraordinary artistic talent into creating the cover design. My appreciation also to my graphic artist Kerry Lampkin who did the many line drawings that have been included to improve the visual presentation of the book. I also want to express my appreciation to Linda Dundas, President of Legal Books Distributing, and to her outstanding staff – Ted Rogers and Mike O'Mahony – for handling the distribution of the book so capably and with so much care and concern.

Finally, I would like to particularly acknowledge my wife Kathy Landis, who wrote the excellent section on Communication Skills in Chapter 6 and who did major editing on this edition of the book. Her gifts as a writer and editor have made the book much easier to read and understand.

Raymond B. Landis
February, 2013

PROLOGUE
What This Book Has to Offer
and How to Get It

It isn't what the book costs. It's what it will cost you if you don't read it.

— Jim Rohn

About two years ago, I was speaking to an *Introduction to Engineering* class at a local university that was using my book. At one point, I asked the students whether they had any questions for me, and one student raised his hand. He asked, "Does the material at the beginning of Chapter 5 on 'Reading for Comprehension' apply to the reading of your book?"

"Yes. Of course!" I replied.

The student then asked,

"So why did I have to wait until I got to Chapter 5 to learn about it?"

I thought this was such a well-taken point that I decided to write this "Prologue" to achieve two objectives:

- **Convince you of the potential of this book (and a course that uses it) to make a difference in your life**
- **Give you guidance on how to realize that potential**

POTENTIAL FOR MAKING A DIFFERENCE

You may be reading this book on your own or as a course requirement. Regardless of the context, it promises to make a significant difference in your life, as both an engineering student and an engineering professional.

Most likely, you have never read a book like this or taken a course like this, and you probably won't again. Most of the courses you take will be about content and the application of that content to solving problems. *Studying Engineering* and a course that uses it are about **you**.

I contend that the maximum potential of this book and a course that uses it to make a difference in your life is far greater than that of any single content-based course you will take. The graph to the right illustrates this contention.

As the graph shows, in most content-based courses you will realize something close to the maximum available potential. If you get an *"A"* grade you probably

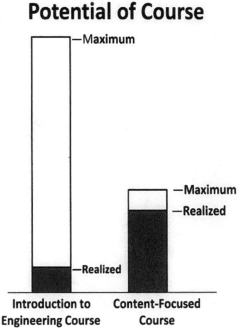

Potential of Course

Introduction to Engineering Course

Content-Focused Course

got 90 percent or more of what was there for you. If you get a *"B"* grade, you probably got 80-90 percent of what was there. And so on.

Unless you are proactive, in spite of the best efforts of your course instructor, you are likely to realize far less than the maximum potential available from this book and much less even than you will get from a single content-based course. *Getting that maximum potential is to a great extent up to you*. And the payoff will be enormous. Not only will you develop academic skills that will enhance your success in engineering study, those skills will correlate closely with the skills you will need to be a successful engineering professional.

I often compare *Studying Engineering* to a mirror. When you get up in the morning, you clean, groom, and dress yourself. And perhaps the last thing you do before going out and confronting the world is to glance into a mirror. Why? It's because you have standards about your appearance, and you are checking to see that these standards are met.

Studying Engineering is like a mirror you look into to access other kinds of personal standards – those beneath your surface appearance. It helps you stand back and reflect and work on characteristics of your deepest self: What kind of person are you? What values do you hold? Do you know what you want out of life, and are you on track to get it? Are you getting the most out of your education? What is your learning process, and how well is it working for you? Based on the insights you derive from the book, you will be prompted to make changes to move closer to the standards you set for yourself.

I hope that you have already committed to a personal goal of receiving your Bachelor of Science degree in engineering. You may not have thought about this explicitly, but achieving a challenging goal like getting your degree in engineering cries out to most students to change.

Change what?

- *The way you think about things (your attitudes, values, mindsets, world views)*

- *The way you go about doing things (your actions, behaviors)*

Indeed, this book and a course that uses it have no value except that they bring about significant changes in you – changes you are aware of and can articulate.

Another way to look at this book is to imagine you are on a merry-go-round and trying to grab a brass ring as you go around. You reach for it, but you miss it. Since the carousel goes around and around, you have more than one chance to grab that ring. This book (and accompanying course) may only "go around once." It may be your one chance to grab the brass ring of growth and development. Don't miss it!

There is some possibility that you are not ready for what this book offers you. How could that be? How could you pass up the opportunity to become a more effective learner and a more successful student? Here's how. You may not pay attention to much of anything, so why would you pay attention to what this book (or class) teaches you? You may not believe that you can change. Or you may not want to change, assuming the way you approached your studies in high school will work in university-level engineering study. Or you may not know what or how to change.

If you aren't open to what this book has to offer, I hope you will keep it in mind should you run into academic difficulty down the line. Rather than give up, thinking you can't be an engineer or you can't get a college education, come back to the principles in this book. They work! I don't believe you can put what's in this book into practice and fail. I truly don't think it's possible.

HOW TO REALIZE THE MAXIMUM POTENTIAL FROM THIS BOOK

If you want to get the most out of the book, I suggest you adopt the following approach [1].

Read a portion of *Studying Engineering* (sentence, paragraph, section, or chapter); then close the book and develop answers to the following four questions:

1. What are the key ideas contained in what you read? (i.e., prepare a summary of what you read either in written form or by saying it out loud).

2. What does the passage you read <u>mean</u> to you? Does it make sense? Does it fit with your way of thinking? Does it fit with your past academic experiences? Are you persuaded of its efficacy? Are you likely to make changes based on it?

3. What questions would you like to ask the author or your instructor?

4. What can or will you change (in either your attitudes or your behaviors) as a result of what you read?

You may need to go back and reread sections to make sure you've gotten everything that's there.

The exercise on the next page will give you an opportunity to try out this approach to learning from this book. Stop and complete this exercise before continuing on.

If you apply the four-step methodology described above to reading *Studying Engineering*, I guarantee that you will come very close to realizing the maximum potential.

I often receive testimonials from students who have taken a course using *Studying Engineering*. An example of such a testimonial is the last section of this Prologue. Read this testimonial (Pages 5-7) with three purposes in mind:

1) To hear from another first-year engineering student.

2) To answer the question: "Could you write such an articulate and insightful statement about your learning process?" If you can, do so. If you can't, set a personal goal to reach the point where you could. This book can help immensely with that.

3) To apply the four suggested steps for reading this text:

   a) Describe what you read.

   b) Describe what you read means to you.

   c) Formulate several questions you would like to ask this student.

   d) Commit to one or more changes you are willing to make in your learning process as a result of reading this testimonial.

ADDITIONAL WAYS TO GET THE MOST FROM THIS BOOK

There are three additional resources that can aid you in getting the most out of this book:

1) The reflections throughout each chapter

2) The problems at the end of each chapter

3) Term Design Project – "Design Your Process of Becoming a 'World-Class' First-Year Engineering Student" (See Appendix A)

If you are taking a course that uses this book, your instructor may assign you the task of developing written responses to some or all of the reflections, to complete some of the end-of-chapter problems, and to do the "Design Your Process" term project. If you are not required to do this work or you are reading the book on your own, I encourage you to do as many of these tasks as possible.

The Reflections. Reflections are interspersed throughout each chapter. The idea is that you read a section and then stop to engage in a guided reflection about what you just read. Thinking about the reflections is good, but writing a response to them is even better.

You will find forms you can use to complete written responses to each reflection in the text at www.discovery-press.com/reflections.htm. Hopefully your instructor will require you to complete them and submit them electronically. But if he or she doesn't, you would still benefit from doing them.

The End-of-Chapter Problems. There is a set of problems at the end of each of the eight chapters of this book. The total number of problems in the entire text is 203. Some of these problems are short and will only take a few minutes, while others will require significant time (e.g., read a biography of a famous engineer and write a critique of the book). I hope you will complete as many of these problems as possible. Doing so will provide a significant learning experience.

If you are taking a course using this book, I expect your instructor will assign you to do a representative number of the problems. If he or she doesn't or you are reading this book on your own, I hope you will step up to the plate and do as many of the problems as possible. Several years ago, a professor told me as an extra credit assignment he had a student do <u>all</u> the problems in the book. I'm sure that student learned a great deal from that exercise, including why not to need to seek an extra credit assignment to pass a course.

Design Project – "Design Your Process." In Appendix A (Pages 276-277), you will find an innovative project – "Design Your Process for Becoming a World-Class Engineering Student." This project could be assigned by your *Introduction to Engineering* course instructor or, if not, you could do it on your own. Through the project you will have the opportunity to compare where you are to where you would need to be to be a "world-class" engineering student and to develop a plan for getting there. Areas covered include strengthening your commitment, utilizing important resources, building relationships, becoming effective at managing time and tasks, adopting appropriate behaviors and attitudes, getting involved in co-curricular activities, and growing through self-assessment.

I wish you the very best in engaging *Studying Engineering*. I always enjoy hearing from students by email (rlandis@calstatela.edu) and make every effort to respond to student comments and questions.

STUDENT TESTIMONIAL (by Nathan Tyson, Messiah College)

There are many strategies that help one achieve academic success. Many of these strategies I have been applying for much of my life, while there are some I have more recently employed, and still others I have yet to put into action. *Intro to Engineering* has made me aware of many flaws in my study practices and has helped me practice a great deal of metacognition. *(Note: Metacognition is discussed on Pages 95-96 in Chapter 3. - RBL)*

Of the many success strategies: getting enough sleep, studying in a non-distractive place, making to-do lists, and asking for help have been weapons in my arsenal for much of my school career.

For all my life, I have been very good about getting enough sleep. If I don't get the proper amount of sleep, I have great difficulty concentrating on my schoolwork and thinking clearly, especially in areas like math and science.

During high school, I spent a lot more time studying than most people. Many of my high school friends would brag about not studying for a single test. I, on the other hand, cannot say the same. In fact, I don't know if I can recall a test that I didn't study for in some way during my high school career.

I learned early on about the importance of studying in a non-distractive place. My mind tends to wander quite easily and is quickly distracted by the most trivial of matters. I cannot study in a room where music is playing, a television is on, or people are talking. In order to study to my full potential, I have found that I need a completely noise-free environment like the library. When I need to study in my room, I have found that earplugs can help immensely by cutting out the distracting background noise.

I have stayed organized by writing to-do lists since my middle school years; however, due to the much heavier college workloads, I have found that simple lists are not adequate enough when it comes to managing my time. In the last few months, my time has become more valuable than ever before; every minute must be spent wisely if I want to succeed in engineering study.

This course has taught me the importance of my time and keeping a schedule, reminding me that time lost can never be retrieved. In mid-October of this year, I started scheduling my week as Dr. Gray [my professor] taught us. Not only have I scheduled every class, but have also scheduled time to complete homework, eat, and sleep. Nearly every hour of every day is planned in order to minimize time wasted and maximize productivity. Since I started using my weekly schedule, I have found that my time seems to have increased exponentially. I feel less overwhelmed by large amounts of work because I know, if I follow my schedule, I *will* get it done.

My schedule has also helped me prepare for tests better than ever before. Back in high school, I would usually save my studying for the night before a test. College, however, is a very different story. I have found that I must start studying *at least* a week before a major test. When I started this, towards the end of September, I was blown away by how much easier and less stressful it was. I can study just an hour a night for a week before a test, and by the time test day rolls around, I have over seven hours of studying and a full night's sleep under my belt. This is much less stressful than staying up all night cramming for a test until I am so stressed out and so tired that I can no longer focus.

Another academic success strategy I have learned from this course is group study. I was always a loner when it came to my studies. Throughout high school, I would do all my work and all my studying by myself. During my first two weeks of college, however, I felt like everyone I talked to was encouraging me to study in groups. I heard it so many times from Dr. Gray, from the *Studying Engineering* book, and from older engineering majors, that I knew I had to try it.

My friends and I now study together on a regular basis. We often work together on calculus assignments, working out problems alone first and then bouncing ideas off each other if we are having difficulty. When one of us understands a difficult problem, he teaches the rest of the group, walking us through it until we all understand. Three of us reviewed together for the first *Intro to Engineering* test and found it quite beneficial. In fact, I believe everyone from our study group got an *"A"* or high *"B"* on the exam. It's amazing what can happen when people get together and teach each other.

Some of my proudest academic moments this semester came as a result of metacognition. Acing every test with little effort is nice, but there is something special about working hard in a tough class and improving. One of my favorite stories of the semester is about my recent improvement in calculus. Earlier in the course, I would do all the homework and pay attention in class, but was always disappointed by my test scores. When I got a 76 percent on the second test, I knew something had to change. The problem was not that I could not do the math, but that I could not do it fast enough. In every test, I would take my time on problems, checking and rechecking my answers, not moving on until I had found the answer. By the time the test was over, I would be left with several incomplete problems and a poor grade. As I began to work on my test taking and study methods, my grade jumped from a 76 to an 88 on the third test. On the fourth and final test of the semester, after much hard work, I received not only my long-awaited *"A,"* but a 100 percent!

REFERENCE

1. Adapted by Anthony Donaldson, Dean of Engineering, California Baptist University from *Teaching Around the 4MAT Cycle: Designing Instruction for Diverse Learners with Diverse Learning Styles,* by Bernice McCarthy and Dennis McCarthy, Corwin Press, Thousand Oaks, CA, 2005.

CHAPTER 1
KEYS TO SUCCESS
IN ENGINEERING STUDY

Success is getting what you want;
happiness is wanting what you get.

— Dale Carnegie

INTRODUCTION

This chapter introduces you to engineering study – both the process that will ensure you succeed and the benefits you will get from doing so.

First, we make our best effort to convince you that you <u>can</u> do it: that success in engineering study, like success in anything you attempt, is a process that you can learn and master just like the many, many other successful students who came before you did.

We point out, however, a mindset that keeps some high-ability, well-prepared students from mastering that process – overconfidence. Students who naively assume that their ability will carry them through engineering study as it did in high school can have a rude awakening.

Next, we discuss two concepts fundamental to success: "goal identification" and "goal clarification." We also emphasize the importance of strongly committing to your goals once you have identified and clarified them.

Then, we present three important keys to success in engineering study:

Effort – *Work hard*
Approach – *Work smart*
Attitude – *Think positively*

As these keys to success suggest, achieving any challenging goal depends largely on your attitudes and behaviors – and for many students that means changing them.

Next, we offer two models to help you understand the skills and knowledge you will get from a quality engineering education, plus a third model to guide you towards obtaining that quality education.

We close the chapter by discussing the need for you to structure your life in ways that will minimize distractions and interferences. Only by doing so will you be able to devote adequate time and energy to your studies and take advantage of the many resources available to you.

The material introduced in this chapter will provide a foundation for you to build on as you study the other chapters of this text.

1.1 YOU CAN DO IT!

From time to time, I meet practicing engineers who tell me about the time when they were first-year engineering students and the dean told the students in their *Introduction to Engineering* class:

> ### *Look to your right; look to your left. Two of the three of you won't be here at graduation.*

It doesn't surprise me that engineering deans (and professors) may say such upsetting things to students. They think that by scaring students about engineering study, the students will be more motivated to succeed.

What does strike me, however, is how angry these practicing engineers are at the dean for having given them such a negative message. And in some cases the event happened some 30 years before! These former students are still upset that the dean tried to frighten them at a time when they were unsure of themselves and easily intimidated.

When I meet with first-year engineering students, I convey a very different message. My message to them and to you is:

Each and every one of you can be successful in graduating with your Bachelor of Science degree in engineering.

How can I make such a bold statement without any specific information about your background or your ability? I'll tell you how.

POORLY PREPARED STUDENTS HAVE SUCCEEDED

For ten years I directed a program designed to enhance the academic success of engineering students. During that period I worked closely with more than 1,000 students. We had students with very poor preparation and limited ability: students who had to take college algebra three times before making a passing grade; students who failed trigonometry and had to repeat it, and then took Calculus I, received a *"D"* and had to repeat it. Some of those students took more than nine years of full-time study to complete their engineering degree.

I ran into one of those students many years later. He was a successful professional engineer and a respected member of his community. When I saw him, he was on the way to drop his daughter off at a relative's home so he could fly to Washington, D.C. for an important meeting.

HIGHLY QUALIFIED STUDENTS HAVE FAILED

I also worked with students who had all the preparation in the world – students who had gone to the best high schools and had excelled in their advanced mathematics and science courses. Yet they did not succeed in engineering study. Some flunked out. Some just dropped out.

The common denominator for such students was that they were <u>overconfident</u>. They had been able to excel in high school without a great deal of effort or a need to adopt effective learning strategies. And they made the mistake of assuming that engineering study would be like high school. They naively believed that their ability would carry them through as it had before. They failed to account for the fact that the faster pace and higher expectations for learning would require substantially more effort and improved learning skills. They didn't recognize that they had moved from the minor leagues to the major leagues, where the students they were competing against had much more ability.

A few of those students have come back to see me. They express their deep regret for not sticking it out. It saddens me to hear they're working in unrewarding jobs for minimum salaries and would like to come back to school, but now the circumstances of their lives prevent them from having a second chance.

I hope you are not such a student. One early indication of this is how receptive you are to the material presented in this book. Thinking there is nothing of value here for you is a sign that you are overconfident. If you are, I hope you will consider this section as a **wake-up call**.

You <u>can</u> ignore this warning with the intent of shifting gears later. The problem with that approach is your early courses, particularly in mathematics and science, provide the foundation on which your entire engineering education will be built. If you start out with a weak foundation, you will find it difficult, if not impossible, to build a sound structure on top of it.

REFLECTION

In the previous two sections, we noted that "poorly prepared students have succeeded" while "highly qualified students have failed." Do you see something of yourself in either category? Do you lack confidence? If so, are you beginning to believe you can do it? Or are you overconfident? If so, are you beginning to become receptive to learning new strategies and approaches for your engineering studies?

WHAT MAKES THE DIFFERENCE?

One student with seemingly limited ability and poor preparation succeeds. Another student with outstanding ability and excellent preparation fails. How can that happen? What are the keys to success in engineering study? What are those things you can do that will virtually ensure your success – those things that, if not done, will at best result in your working below potential and even lead to failure?

Success in engineering study is not unlike success in anything you have attempted or will attempt. Achieving success is a process, and each step in the process can be learned. I would encourage you to make a commitment to become an "expert" on success. It's something that you can do. And the payoff will be enormous.

Lots of resources are available to help you. You can learn from others, from reading books such as Stephen Covey's *The Seven Habits of Highly Effective People* [1], listening to audio CDs, tapping the wealth of information available on the Internet, and attending short courses and workshops.

The Internet is an easy, reliable guide for identifying the resources that best speak to you.

For online reading material, Google "success." If the results are overwhelming, try narrowing your search to "student success strategies" or even "engineering student success strategies." Play around with keywords until you hit on the most promising results.

For books and CDs, www.amazon.com is golden. Under "books" enter the keyword "success." To locate audio CDs, click on "audiobooks," followed by the keyword "success."

You'll find videos and lectures about success on www.YouTube.com and www.ted.com. Go to one of these websites and conduct a search on the keyword "success."

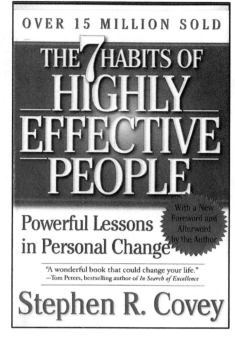

Then too, don't ignore the many resources listed at the end of each chapter of this book. Make learning about success one of your life goals. If you work at it, your capacity to be successful will expand and grow. You might even surprise yourself at what you can achieve.

And who knows? Maybe someday you'll write a book on *success* for others. When I was sitting in your seat, I could never have imagined I would someday write a book like this one.

1.2 WHAT IS "SUCCESS"?

I assume you want to be successful. I hope that's why you are reading this book. But just wanting to be successful is not enough. Everyone wants to be successful. Often when speaking to an *Introduction to Engineering* class, I'll ask the question, "How many of you want to be successful?" All of the students raise their hands. But what do the students mean when they indicate they want to be successful? Are they all thinking about the same thing? Probably not.

When I ask the same students, "What is *success*?" I get a variety of answers:

➤ *Success* is being happy.

➤ *Success* is making money.

➤ *Success* is having control over your life.

But almost always one or more students will give the correct answer:

➤ *Success* is the achievement of goals.

Webster's Dictionary says essentially the same thing:

<u>*Success*</u> *is the achievement of something desired, planned, or attempted.*

The point is that unless you have something "desired, planned, or attempted," there can be no success.

Unfortunately, many students lack a clear goal and commitment to that goal necessary for success. According to Vincent Tinto, author of an excellent book on student success [2], the top two reasons that students do not succeed in college are:

(1) Lack of Intention – Students do not have a clear educational and/or career goal.

(2) Lack of Commitment – Students do not have the motivation and drive to work toward attaining their educational/career goals.

Identifying a clear goal and developing a strong commitment to that goal are the essential first two steps in the process of achieving success.

REFLECTION

Reflect on the relationship between *success* and *happiness*. What does each of these words mean to you? Does success bring happiness? Can people be happy if they are not successful? Think about Dale Carnegie's quote at the beginning of this chapter: *Success is getting what you want; happiness is wanting what you get.* Do you usually get what you want? Do you usually want what you get? What insights can you derive by contemplating the relationship between <u>*success*</u> and <u>*happiness*</u>?

GOAL SETTING

If success requires a goal, let's discuss goal setting. Obvious though it may sound, the basic idea behind goal setting is:

How can you ever expect to get somewhere if you don't know where you want to go?

That is, setting goals – having specific ideas of what you want to accomplish in the short and long term – is a key requirement to becoming an effective student and professional. Only when you set goals will you have something to strive for and something against which to measure your progress.

GOALS GIVE YOU SOMETHING TO MEASURE YOUR-SELF AGAINST.
Consider, for example, two engineering students in a calculus class who score a "*B*" on their first exam. One student is extremely

unhappy and resolves to study much harder for the next exam. She has set a goal of earning an "*A*" in the course and by falling short on the first exam, she knows that she must work more. The other student, however, is content with the "*B*" grade and so decides he can increase his outside work hours since even less study is necessary than he thought.

These different responses results from the different expectations these two students have , based on their goals. This illustrates how success or failure can only be measured according to self-imposed goals.

Goals Give Your Life Direction. I'm sure you were asked many times during your childhood, "What do you want to be when you grow up?" If you didn't know, you probably felt a bit frustrated and even irritated at people who asked you that question.

I hope you realize by now that they were trying to help you. They were trying to alert you to the importance of setting directions for your life. They probably even realized intuitively that the more reluctant you were to grapple with this question, the more important it was that <u>you</u> of all people do so.

Setting goals may not be easy, but the payoff is definitely worth the effort, as the stories of many successful people indicate. Following is but one such story.

Astronaut Franklin Chang-Diaz

Dr. Franklin Chang-Diaz is one of the most accomplished astronauts at NASA. A veteran of six space missions, he has logged nearly 1,300 hours in space. But when you hear the story of his life, you wouldn't think he'd end up in such a prestigious position.

Chang-Diaz was born and raised in Costa Rica. As a child he was enamored by U.S. space program. He and his friends used to build spacecrafts out of cardboard boxes, equipping them with broken radios, furniture, and other discarded material. They would then go through a countdown and lift-off and pretend to travel to distant planets.

Franklin Chang-Diaz

Because of his interest, Chang-Diaz set a personal goal of becoming a U.S. astronaut. Imagine a young Costa Rican citizen who didn't speak a word of English aspiring to be a U.S. astronaut!

When he finished high school, he worked for a year and saved enough money to buy a one-way airplane ticket to Hartford, Connecticut, where he had some distant relatives. In Hartford he repeated his senior year of high school, learned English, and was admitted to the University of Connecticut, where he majored in engineering. After graduating with honors, he began graduate study at MIT, eventually receiving his Ph.D. in plasma physics. He then applied for the astronaut program, was accepted, and became the U.S.'s first Hispanic astronaut.

To learn more about Dr. Chang-Diaz and his career as a U.S. astronaut, visit his NASA website at: www.jsc.nasa.gov/Bios/htmlbios/chang.html.

The point that the story of Dr. Chang-Diaz drives home so convincingly is the need to have goals. His story makes me wonder what I might have accomplished had I set such lofty goals.

<u>**WRITE DOWN YOUR GOALS.**</u> Right now your primary goal should be to graduate with your degree in engineering. But what else would you like to accomplish? Become president of your own company? Become a multimillionaire? Become a college professor? And what about your more immediate goals? Maybe you want to make a 3.0 GPA next term, improve your writing skills, or become president of one of the engineering student organizations.

A good exercise would be for you to write down your short-term, intermediate-term, and long-term goals. Consider what you want to accomplish in the next week, in the next month, in the next year, in the next five years. Review and update these lists regularly.

Start by making graduation in engineering one of your primary life goals.

REFLECTION

Think about how goals, dreams, and fantasies differ. What distinguishes each? Do you have goals? Dreams? Fantasies? What does it take to convert a dream into a goal?

STRENGTHENING YOUR COMMITMENT

Why did you choose engineering as your major? Perhaps because you were good in math and science, one of your high school teachers or counselors recommended that you study it. Or maybe you are doing it to please your parents, or you don't know what else to study. It is likely that you don't know a great deal about engineering. Few students do.

Regardless of your reasons for electing engineering, it is critically important that you develop a strong motivation to succeed. Engineering is a demanding field of study. Even a student with excellent preparation and strong ability will not succeed without a high level of commitment.

There are at least three practical strategies you can use to strengthen your commitment to success in engineering study:

    (1) Clarifying your goals

    (2) Learning as much as you can about engineering

    (3) Developing a "road map"

<u>**CLARIFYING YOUR GOALS.**</u> What does it mean to clarify your goals? Very simply, it means answering such questions as, "Why do I want to achieve the goal?" "What will the payoff be?"

"What will it mean to the quality of my life if I am successful in achieving the goal?" Clarifying your goals helps you understand their value to you. And by better understanding their value, you will become more committed to achieving them.

As noted earlier, many students know very little about engineering and what engineers do. In particular, they tend not to know about the tremendous rewards and opportunities that an engineering degree offers. Learning about these rewards and opportunities, as we will do in Chapter 2, will figure significantly into clarifying your personal goals.

LEARNING AS MUCH AS YOU CAN ABOUT ENGINEERING. As you have grown up, you have been exposed to teachers, doctors, dentists, and numerous other professionals. You have a feel for what accountants do if you have had to manage your personal finances. You have seen lawyers at work on TV shows such as *Law and Order*. Through your coursework, you have developed some feel for what mathematicians, chemists, and physicists do. It is doubtful, however, that you have had much exposure to engineering. The exposure you have had has probably been indirect, through contact with the products that engineers design.

Learning about engineering is a lifelong process, but it should begin now. Take advantage of every opportunity that presents itself. You can start by studying Chapter 2 of this text thoroughly. Explore some of the many Internet websites referred to there, particularly those whose purpose is to help students learn about engineering. Attend seminars on career opportunities, go on field trips to industry, and talk with company representatives at career day programs. Browse the resource library in your career center. Become active in the student chapter of the professional engineering society for your major. Talk to your professors. Read biographies of successful engineers [3, 4, 5, 6, 7]. If you land a summer job in industry, be curious and inquisitive. Look around. Talk to the engineers there and find out what they do.

Over time, these efforts will pay off and your understanding of engineering will increase. Increased knowledge brings increased motivation. We tend to like things we know a lot about.

PREPARING A ROAD MAP. Remember when you were in elementary school and heard the term "algebra" and thought, "I'll never be able to learn that!" And later you were overwhelmed with the thought of mastering trigonometry or calculus. You assumed you wouldn't be able to handle such advanced subjects, but you were wrong. Each time you reached the next higher level, you were able to handle it, even excel at it. How did you do it? By taking lots of little steps, each one building on previous steps.

Often students ask me, "What does it take to succeed in engineering study?" My answer is, "You must be able to pass Calculus I at the university level." My reason for this is very simple. If you can pass Calculus I, you can pass Calculus II. And if you can pass Calculus II, you can pass Calculus III. If you can pass Calculus III, you can then pass Calculus IV. And if you can pass these calculus requirements, you can pass the junior engineering courses. If you can pass the junior engineering courses, you can pass the senior engineering courses.

So you see, succeeding in your engineering program is a process of taking one little step after another. Progressing through the engineering curriculum is just an extension of what you have already demonstrated you can do.

I suggest you develop a "road map" that will lead you to graduation in engineering. Lay out a plan of what you will need to take each semester or quarter. Having a step-by-step road map to follow will increase your confidence and strengthen your commitment to achieve your ultimate goal: that B.S. degree in engineering.

DON'T LET ADVERSITY STOP YOU. Highly successful football coach, ESPN sports analyst, and motivational speaker Lou Holtz notes a primary difference between people who succeed and people who fail.

Lou Holtz

> *People who succeed are people who, when they get knocked down by some adversity, get up; whereas, people who fail are people who, when they get knocked down, stay down.*

I would encourage you to read Coach Holtz's autobiography *Wins, Losses, and Lessons* [8]. Or you can hear from Coach Holtz by watching his 14-minute University of Portland 2012 commencement speech at www.youtube.com/watch?v=nkR0iVEcDEE.

The most likely reason you will fail to graduate in engineering is that you will encounter adversity and give up. You will have difficulty with a course or a professor. You might have a personal problem. Whatever adversity you are bound to experience, you will be tempted to use it as an excuse or justification for quitting. DON'T!

By strengthening your commitment following the steps outlined in the previous three sections, you will develop determination. The dictionary defines determination as "a firmness of purpose . . . [and] having one's mind made up." Determination means having an unwavering commitment to your goal: the goal of graduating in engineering. You must be determined to persist, particularly in the face of adversity.

A Personal Story

I dropped out of college early in my sophomore year. When I attempted to register for my second year, I learned I had lost my full tuition scholarship because of poor grades. Faced with having to take out a massive student loan and having broken my leg playing intramural football, I dropped out.

I had always wanted to be a jet pilot, so as soon as my leg healed, I went directly to the local Air Force Recruiting Office. To my chagrin I was told a college degree was required for acceptance into flight training. Soon I was back in school with newfound determination. That experience was a significant lesson to me that doors would be shut without a college education.

Adopt the view that you are going to achieve your goal and that nothing is going to stop you. And how do you keep adversity from stopping you? How can you keep failures from discouraging you? I find this age-old saying to be very helpful as a philosophical basis for overcoming adversity:

We learn more from our failures than we do from our successes.

It's true! Think about it.

Another Personal Story

When I was in the 7th grade, I took a gymnastics class. I was the best in the class on the pommel horse. So when we had a competition at the end of the term, everyone expected I would win that event. But when I began performing, I was so nervous I felt as if needles were pricking my skin all over. I came in last place. I was terribly embarrassed and ashamed. It took me a long time to get over that failure. But that experience showed me that if I take myself too seriously and want to win too much, I can actually perform much worse than I am capable of. That experience has helped me deal effectively with high-pressure situations ever since.

Learning to overcome adversity as a student will also benefit you during your professional career. Joseph J. Jacobs, founder and former CEO of Jacobs Engineering and one of the nation's most successful engineers, gives his "Nine Commandments for the Entrepreneur." The first four are:

(1) You must be willing to risk failure.

(2) You must passionately hate failure.

(3) Persistence is a necessity, as is the willingness to acknowledge defeat and to move on.

(4) A measure of your potential to succeed is how you handle adversity.

(I encourage you to read Mr. Jacobs' highly motivational autobiography, *The Anatomy of an Entrepreneur* [6].)

If you are determined to graduate in engineering, if you persist even in the face of adversity, if you take the view that you will not allow anything to stop you, the chances are very good that you will succeed.

Believe in yourself. You can do it!

REFLECTION

What do you think about the claim that "You learn more from your failures than you do from your successes"? Have you ever experienced a significant failure? What was it? What did you learn from that experience?

1.3 KEYS TO SUCCESS IN ENGINEERING STUDY

Setting a goal and making it important to you are only the first steps. The real challenge remains – achieving the goal. Once your goal is identified and you have done everything you can to develop a strong commitment to that goal, achieving it requires that you adjust both your attitudes and your behaviors. This means that you base your day-to-day decisions and choices on whether a particular action supports your goal (i.e., moves you closer to it) or conflicts with your goal (i.e., moves you farther away from it). The same applies to attitudes you hold.

In my experience there are three keys to success in engineering study:

Effort - *Work hard*
Approach - *Work smart*
Attitude - *Think positively*

Let's examine each of these.

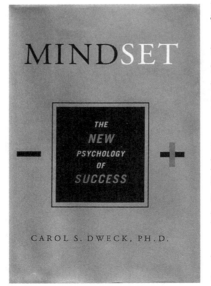

EFFORT – "WORK HARD"

Do you believe that people succeed because of their **ability** – that some people "have it" while others don't? Or do you believe that people succeed because of their **effort**? An excellent book that contrasts these two ways of looking at the world is *Mindset: The New Psychology of Success* by Carol S. Dweck [10].

The first belief – that some people have it and some don't – is self-defeating. Thinking you don't have as much ability as others provides a rationale for you to accept personal failures: You may as well give up. After all, if success is related to some natural quality that you have no control over, then it doesn't matter what you do or how hard you work.

Believing you're the smartest kid on the block has pitfalls as well. If you do, you're likely to feel the need to prove yourself over and over, trying to look smart and talented at all costs.

Research has shown that people with this fixed mindset are more likely to stick with approaches that clearly don't work, while ignoring suggestions from others.

The second belief – that people succeed because of their effort – is empowering because the amount of effort you put in is in your direct control. You can choose to put in more effort and in doing so significantly affect your success.

The following table compares how people with fixed mindsets view challenges, obstacles, effort, criticism, and the success of others with how people with a growth mindset view the same items.

	FIXED MINDSET	GROWTH MINDSET
Challenges	Avoid challenges; stick to what you know well	Embrace challenges; overcoming challenges makes you stronger, smarter
Obstacles	Give up easily when you encounter obstacles	Persist in the face of setbacks; failure is an opportunity to learn
Effort	See effort as unpleasant and fruitless; the need for effort is a sign of low ability	See effort as the path to growth and mastery
Criticism	Ignore useful criticism; see negative feedback as an insult	Seek feedback and learn from criticism
Success of Others	Feel threatened by the success of others	Find lessons and inspiration in the success of others

> **REFLECTION**
>
> Would you say you tend to have a fixed mindset or growth mindset? Think about the way you deal with challenges and obstacles. What is your view about effort? How do you deal with criticism? How do you view other peoples' success? Which mindset do you think would bring a person more success and more happiness in life? If you see some of the fixed mindset traits in yourself, what could you do to change your mindset?

ABILITY VS. EFFORT. The relative importance of ability and effort was perhaps best explained by the famous American inventor Thomas Edison:

Genius is one percent inspiration and 99 percent perspiration.

Does the following dialogue sound familiar to you? Over the years, I've had a variation of it with many of my students.

Landis: How's everything going?

Student: *Fine!*

Landis: What's your hardest course this term?

Student: *Physics: Electricity and Magnetism.*

Landis: How are you doing in that course?

Student: *Fine!*

Landis: What score did you make on the last exam?

Student: *Forty-three.*

Landis: What grade is that?

Student: *I don't know.*

Landis: Is it an *"A"*?

Student: *No.*

Landis: A *"B"*?

Student: *No.*

Landis: A *"C"*?

Student: *Probably not.*

Landis: A *"D"*?

Student: *Maybe.*

Landis: 'Sounds like an *"F"* to me. How many hours are you putting into your physics course?

Student: *About 15 hours a week.*

Landis: How many hours have you studied today?

Student: *I haven't done any studying today.*

Landis: How many hours did you study yesterday?

Student: *None.*

Landis: How about over the weekend?

Student: *I meant to, but just never got to it.*

Landis: So you're planning to study physics for five hours a day for the next three days to get your 15 hours in this week?

EFFORT IS BOTH TIME AND ENERGY. In my experience, poor academic performance can usually be traced to insufficient effort. Just what do I mean by "effort"? It is using energy, particularly mental power, to get something done.

The effort you devote to your studies has two components: time and energy. An analogy can be made using the well-known physics formula:

Distance = Rate x Time

Completing a specific task (distance) requires that you devote energy or mental power (rate) and spend time on the task (time). In later sections, we will consider how much time is sufficient, what is the best use of that time, and when to put in that time if you want to be an effective and efficient student.

The important point here is that your success in studying engineering is largely **in your control**. How well you perform will depend, in large measure, on how much effort you put in. Accomplishing an academic task, like completing a homework assignment, will require you to devote adequate time and to focus your mental energy.

These are things that you can choose to do or choose not do.

APPROACH – "WORK SMART"

"Approach" refers to *how* you go about your engineering studies. It means that you work not only hard but smart. In large measure, your approach to your engineering studies depends on the ideas we have already discussed. It assumes that:

- You know why you want to be an engineer and appreciate the value of a technical education.

- You have clarified your goals and developed a road map to lead you to them.

- You have a strong commitment to achieving your goals, even in the face of adversity.

- You have gotten your life situation in order, so that you are not overburdened with problems and distractions.

Above all, however, your approach to your engineering studies – working "smart"– means that you learn to become a *master* engineering student.

BECOMING A MASTER STUDENT. To understand what I mean by becoming a *master* student, consider the following analogy. If you were to take up chess, what would you do? Learn the basic objectives, rules, and moves and then begin to play? Probably. But you'd soon discover that mastering a game of skill like chess requires much more. So you might read a book, take a lesson, or watch experts play. You would realize that to become a chess *master*, you need to spend time both playing the game <u>and</u> learning how to play it.

Your approach to the study of engineering can be likened to a game. To become a *master* student, you must not only play the game (i.e., be a student); you must also devote time and energy to learning how to play it (i.e., learn to excel as a student).

KLAMPKIN
2012

The first step in playing the game of becoming a master engineering student is to get a clear picture of what is required to earn your B.S. degree. Earlier, when discussing what it means to prepare a road map for yourself, I gave a brief synopsis of what you need to do to graduate in engineering. Let me give you a related description here.

You become an engineer when you pass a set of courses required for an engineering degree. What is required to pass each course in the set? Primarily passing a series of tests and exams. To pass the series of tests, you must pass each test one at a time. So by breaking down the education process this way, you can see that to become an engineer, you must become a master at preparing for and taking tests.

Of course, this is easier said than done, because many other factors are involved. But by approaching your engineering studies in this light, the "game" of becoming a master student and, ultimately, earning your engineering degree becomes less daunting.

As you read the subsequent chapters in this book, you will discover different ideas and perspectives on how best to approach your studies.

Learning to be a master engineering student will be a tremendously rewarding and beneficial experience. It will enhance your immediate success as a student, while developing important skills you will later need as a practicing professional engineer. Indeed, many of the approaches you learn in this book will work for you in whatever you do in your life.

ATTITUDE – "THINK POSITIVELY"

Are you a positive or negative person? Are you aware of the role attitude plays in your success? What do you think of the following statement?

> ### *Positive attitudes produce positive results.*
> ### *Negative attitudes produce negative results.*

Among those negative attitudes that could produce negative results in engineering study are:

- Weak commitment to the goal of graduating in engineering

- Low self-confidence

- Unrealistic view of what's expected to succeed in engineering studies (overconfidence, naiveté)

- Lack of self-worth (leading to tendency to sabotage your success)

- External "locus-of-control" (i.e., adopting a "victim" role)

- Unwillingness to seek help (thinking that seeking help is a sign of weakness)

- Resistance to change your behaviors and attitudes

- Tendency to procrastinate (having a negative view about the idea of managing your time)

- Avoidance of areas of weakness or perceived unpleasantness (e.g., writing, oral presentations, difficult courses)

- Reluctance to study with other students

- Negative view toward authority figures (parents, professors, engineering professionals)

REFLECTION

Think about each of the negative attitudes in the list above. Do any of the items describe you? If so, in what ways could you see that particular attitude interfering with your success in engineering study? Do you know why you hold this attitude? Are you willing to try to change the attitude? What would be a more positive attitude that you could adopt?

One of the primary purposes of this book is to help you become conscious of and change any negative attitudes you may hold that will impede your success in engineering study. You will learn the process for this change when you study Chapter 6: *Personal Growth and Student Development*.

SUMMARY OF THE SUCCESS PROCESS

In the previous sections we have discussed the "success process." Below is a summary of this four-step process as it applies to the goal of becoming an engineer.

Step 1: Setting goals	Do I want to be an engineer?
Step 2: Strengthening commitment to goals	How important is it to me to become an engineer?
Step 3: Changing negative attitudes	What attitudes will interfere with my goal of becoming an engineer?
Step 4: Changing non-productive behaviors	What do I need to do differently to achieve my goal of becoming an engineer?

This book will help you navigate this process.

➢ Chapter 2 will help you firm up your goal of becoming an engineer and deciding which discipline to specialize in.

➢ Chapter 2 will help you strengthen your commitment to becoming an engineer by providing knowledge about the field of engineering, while exposing you to the rewards and opportunities of an engineering career.

➢ Chapter 6 will aid you in the process of changing any negative attitudes to positive ones appropriate to success in math/ science/engineering coursework.

➢ Chapters 3, 4, and 5 will help you adopt the behaviors that will ensure you are studying "smart."

You can choose to devote time and energy to each step in the success process. For example, you might schedule Saturdays from noon to 2:00 p.m. to work on strengthening your commitment to success in engineering study. Or you could devote another block of time to figuring out what behaviors you need to change to be a more effective student.

1.4   MODELS FOR VIEWING YOUR EDUCATION

One of the most positive and unique aspects of your college experience is that you are working for yourself to prepare yourself for your future. Consider the saying:

No deposit, no return.

Your education represents a major deposit, or investment, you are making in yourself. Your return will be in direct relation to what you put in. You must realize that whenever you take the easiest instructor, avoid a tough course, or cut a class, you are hurting yourself. Whenever you make a conscious choice to avoid learning, growing, or developing, you are not getting away with something: **You are working against yourself.**

Having a model from which to view your education will assist you in getting the most out of it. Earlier in this chapter, I gave simplified explanations of the engineering curriculum in order to demystify it for you. First, I described it as a required set of courses you must take. Later, I broke down each course as a series of exams you must pass.

It is time now to broaden your view of your engineering studies because a quality education involves much more.

The purpose of the next three sections is to give you three models from which to view your education. These models will assist you in answering such important questions as:

- What is the purpose of my education?
- What should I know when I graduate?
- How do I know if I am getting an excellent education?
- How can I enhance the quality of my education?
- Will I have the knowledge and skills to get a job and be a successful engineering professional?

These models are also useful for all kinds of personal assessment or self-evaluation. My suggestion is that you measure yourself against each item presented in these models. In other words, ask yourself on a scale of zero to ten (ten being highest): *How would I rate myself on this item?* In areas you feel you are strong, just keep doing what you have been doing. In areas you need to improve, map out a plan to strengthen them. Personal assessment and development plans will be discussed in more detail in Chapter 6.

ATTRIBUTES MODEL

In today's tight fiscal climate, universities are being held more accountable for their productivity. Institutions are being asked to establish educational objectives (desired results) and student outcomes (achieved results), demonstrate how they plan to achieve these objectives and outcomes. This process is called *institutional assessment*. It is not unlike what happens to you in your classes. Your professor sets course objectives and has expectations of how well you should do in achieving these objectives. At the end of the term, the degree to which you meet these expectations is measured and translated in the form of a final grade.

One way engineering programs are held accountable is through the accreditation process administered by the Accreditation Board for Engineering and Technology (ABET). Understanding the accreditation process (discussed in more detail in Chapter 8) will help you better understand the engineering education you are receiving.

ABET, through its *Engineering Criteria 2000* [11], mandates that engineering programs must demonstrate their graduates have the following attributes:

ABET Attributes of Engineering Graduates

a. An ability to apply knowledge of mathematics, science, and engineering

b. An ability to design and conduct experiments, as well as to analyze and interpret data

c. An ability to design a system, component, or process to meet desired needs within realistic constraints such as economic, environmental, social, political, ethical, health and safety, manufacturability, and sustainability

d. An ability to function on multi-disciplinary teams

e. An ability to identify, formulate, and solve engineering problems

f. An understanding of professional and ethical responsibility

g. An ability to communicate effectively

h. The broad education necessary to understand the impact of engineering solutions in a global, economic, environmental, and societal context

i. A recognition of the need for, and an ability to engage in, lifelong learning

j. A knowledge of contemporary issues

k. An ability to use the techniques, skills, and modern engineering tools necessary for engineering practice

This list of attributes provides a clear picture of what you should get from your engineering education. That is, when you complete your engineering degree, you will have the knowledge, skills, and attitudes you will need for a successful and rewarding career.

REFLECTION

Reflect on each of the attributes of engineering graduates required by ABET. Think about what is meant by each item on the list. Reflect on the perspective that this is a type of blueprint for what you should gain from your engineering education. Consider which areas are most appealing to you. Which areas are you likely to excel in? Design? Communication skills? Teamwork? Problem solving? Use of engineering tools? Experimentation? Ethical responsibility? Other?

EMPLOYMENT MODEL

A second model that may be useful to you in viewing your education is the Employment Model. Certainly, one reason why many students choose to major in engineering is the availability of jobs. In light of this, you need to consider what characteristics are important to employers and work to develop yourself in these areas.

An excellent recent study [12] conducted by the Corporate Member Council of the American Society of Engineering Education ranked the following attributes in order of importance for early-career engineering professionals:

1. Communicates effectively in a variety of different ways, methods, and media

2. Possesses the ability to think both critically and creatively

3. Shows initiative and demonstrates a willingness to learn

4. Functions effectively on a team

5. Possesses the ability to think both individually and cooperatively

6. Demonstrates an understanding of engineering, science, and mathematics fundamentals

7. Demonstrates an understanding of information technology, digital competency, and information literacy

8. Maintains a positive self-image and possesses positive self-confidence

Note that the top-ranked items are personal traits (communication skills, initiative, willingness to learn, critical and creative thinking skills, self-image and confidence, etc.). Technical knowledge, including understanding of math, science, engineering fundamentals and information technology, are ranked #6 and #7.

As you approach graduation, you will undoubtedly participate in a number of interviews with prospective employers. How you fare in those interviews will depend largely on how well you have prepared yourself in the eight areas listed above.

Subsequent chapters in this book offer guidance and suggestions to help you acquire these attributes.

- **Chapters 3, 4, and 5** will address academic success strategies that will ensure you have strong scholastic qualifications.

- **Chapter 6** will instruct you in ways to develop your personal qualifications.

- **Chapter 7** will explain the value of active involvement in student organizations and engineering-related work experience.

STUDENT INVOLVEMENT MODEL

Let's assume that you want to get a *quality education* – i.e., to acquire the knowledge, skills, and attitudes that will result in your being highly sought after by engineering employers. How can you guarantee that you get that *quality education*? In fact, what do we mean by "quality" or "excellence" in education? We can find the answer in a classic paper by Alexander W. Astin titled "Involvement: The Cornerstone of Excellence" [13].

GETTING AN EXCELLENT EDUCATION. According to Astin, an "excellent" education is one that maximizes students' intellectual and personal development. He says the key to students' intellectual and personal development is a high level of "student involvement." Astin defines student involvement as:

the amount of physical and psychological energy
that the student devotes to the academic experience.

He gives five measures of student involvement:

(1) Time and energy devoted to studying

(2) Time spent on campus

(3) Participation in student organizations

(4) Interaction with faculty members

(5) Interaction with other students

Put simply by Astin:

> A **_highly involved student_** is one who, for example, devotes considerable energy to studying, spends a lot of time on campus, participates actively in student organizations, and interacts frequently with faculty members and other students.

Conversely, according to Astin:

> An **_uninvolved student_** may neglect studies, spend little time on campus, abstain from extracurricular activities, and have little contact with faculty members or other students.

REFLECTION

Evaluate yourself based on Astin's five measures of student involvement. Would you describe yourself as a "highly involved student"? Or would you describe yourself as an "uninvolved student"? Do you agree that Astin's five measures relate to the quality of the education you are receiving? Are you willing to make changes to take advantage of Astin's model?

INVOLVEMENT IS UP TO YOU. The Astin "student involvement" model suggests that the quality of the education you get will depend primarily on the approach you take to your studies. Although your institution can do things to encourage you to study more, spend more time on campus, become involved in student organizations, and interact with your professors and fellow students, **_increasing your level of involvement is mostly up to you_**. Don't leave it to chance. Be proactive!

You can <u>choose</u> to devote more time and energy to your studies, spend more time on campus, and become active in student organizations. You can <u>choose</u> to interact more with your professors and become more involved with other students. In doing so, you will greatly enhance the quality of your education.

1.5 STRUCTURE YOUR LIFE SITUATION

I hope that the ideas presented thus far in this chapter have convinced you of the importance of making success in engineering study one of your primary life goals and have strengthened your commitment to that goal. One of the key objectives of Chapter 2 is to strengthen that commitment even more by increasing your understanding of engineering as a profession and giving you a clear picture of the rewards and opportunities that an engineering career offers you. With a clear goal and strong commitment to it, you are well on your way to achieving it. All that remains is to do it.

The first step in "doing it" is to create a life situation that supports your goal. Full-time engineering study is a major commitment, so you must be prepared to devote most of your time and energy to it. This means eliminating or minimizing any distractions or obligations that will interfere with your studies and work against your achieving your goal.

I often encounter students who are taking a full load of math/science/engineering courses while commuting over an hour each way to school, working 20 or more hours per week, meeting demands placed on them by their family, and trying to maintain an active social life. Students in such situations are very likely destined for failure.

Whether demands outside of school come from family, friends, work, or commuting, you need to make whatever changes are necessary so that you, too, don't program yourself for failure.

LIVING ARRANGEMENTS

If at all possible, live on or near your college or university campus. The more immersed you can get in the academic environment of your institution, the better your chances of success will be. Commuting takes time, energy, and money; and living at home can present problems. Parents may expect you to help with the household duties. Little brothers and sisters may be noisy and distracting. Neighborhood friends may not understand your need to put your studies ahead of them. Wherever you live, however, remember that now is a time in your life when it's appropriate to be a bit selfish. Place a high value on your time and learn to say no when necessary.

Regardless of your living arrangement – at home with parents, in an apartment alone or with a roommate, or in an on-campus residence hall – I would suggest that you come to campus early in the day and do your work there, rather than coming only to take classes and leaving as soon as possible. Your university or college campus is an academic place. Its primary purpose is to facilitate the teaching/learning process. And it's set up to do just that. Whereas, at home, apartment, or residence hall there are many distractions (TV, DVR, refrigerator, friends, parents, siblings), on campus there are lots of resources (professors, other students, places to study, library resources, tutors).

I would encourage you to approach your engineering study much as you would a full-time job in that you go to your "place of work" and do the greater share of your work there, perhaps bringing some work home, but certainly not all.

Sometimes you'll hear the viewpoint that students at residential campuses get more out of their education than students at commuter campuses. Putting the approach outlined in the previous paragraph into operation, in effect, brings all the benefits of a residential student to a commuting student.

PART-TIME WORK

As noted above, full-time engineering study is a full-time commitment. Working up to ten hours a week at a part-time job is probably okay, but more is almost certain to take its toll on your academic performance. While it may be essential for you to work, it may also be that you are working to afford a nice car, expensive clothes, or other non-essentials. Look at it this way. You may get a job for $8-10 an hour now, but in doing so you will jeopardize your education or at best extend the time to graduation.

The average starting salary for engineering graduates is around $30 an hour ($60,000/year). If your career is successful, someday you might make more per hour than students make per day. So try to delay as many material wants and minimize family obligations as possible. By doing so, you will have much more in the long term.

If you must work while going to school, particularly if your work exceeds ten hours per week, how can you achieve a reasonable balance between working and studying? One way is to follow the guidelines below.

Hours worked	Course load
10 hrs/wk	full load
20 hrs/wk	12 units
40 hrs/wk	8 units

Another way to manage your study and workload is to follow the "60-Hour Rule" espoused by Dr. Tom Mulinazzi, Associate Dean of Engineering at the University of Kansas [14]. His is not a rigid rule, but rather a guideline. It doesn't apply to a single week but is a pretty good rule-of-thumb over the long haul.

The 60-Hour Rule (according to Dean Tom Mulinazzi)

The "60-Hour Rule" is an excellent guideline to follow. I have shared it with first-year students in an *Introduction to Engineering* course each fall.

The Rule assumes that a student can work for 60 hours a week over the period of a term. This work includes academic work, work at a paying job, and time spent commuting. It also assumes that a student studies two hours for every hour spent in the classroom.

Let's say that a student is working 20 hours on campus. Subtract 20 from 60 and the result is 40. Divide 40 by three (one hour in class and two hours studying per week for every credit hour) and the result is 13. This means that most students can take 13 units of coursework and derive satisfactory results while working 20 hours. Ninety-five percent of the engineering students who are dismissed from the University of Kansas, School of Engineering, violate the 60-Hour Rule.

From time to time, I encounter a student who is taking four courses each term but passing only two of them. When I suggest that the student reduce his or her course load, the typical response is, "I can't do that. It'll take me forever to graduate!" Obviously, though, such students are moving through the curriculum at the rate of only two courses per term. The point is, be realistic about your situation. Don't create an unmanageable workload and then deceive yourself into thinking it is working.

INFLUENCE OF FAMILY AND FRIENDS

Because family member and friends may not understand the demands of engineering study, they may unintentionally distract you. If your family members pose problems, have a frank talk with them. Let them know that you want to make school your number-one priority. Ask for their help and negotiate clear agreements about their demands on you.

If you are a recent high school graduate, dealing with friends from high school – especially those who are not pursuing a college education – may be difficult. These friends may put pressure on you to spend as much time with them as you did in high school, while you may find you not only don't have time for them, you also have less and less in common with them.

If you find yourself in this situation, you alone will have to decide how to handle it, as there are no easy answers. But it is important to be realistic and understand the consequences of your choice to study engineering. By making this choice, you are moving yourself in a different direction that may increasingly distance you from your old friends, while bringing you into contact with new people and peers – and opportunities for new friendships.

However you decide to deal with your old friends, by all means do not let them keep you from the opportunities to develop new friendships at school.

I can't encourage you enough to cultivate relationships with your fellow engineering students, for befriending them will be tremendously rewarding. Not only will you likely be initiating important lifelong relationships, you also will derive the immediate benefits of being able to integrate your academic and social lives, while building a support system in which "friends help friends" achieve the same academic goals. If you've watched the TV series *Community*, you know what I'm talking about.

REFLECTION

Think about your life situation in terms of the factors presented in the previous sections. Are there things that come to mind about your living arrangements, workload, or expectations/demands of family and friends that need to be changed if you are going to be successful in engineering study? List those things and come up with ways to bring about the needed changes.

SUMMARY

This chapter introduced you to the keys to success in engineering study. We first focused on the importance of making graduation in engineering your primary goal.

Next we presented three strategies for strengthening your commitment to that goal: (1) clarifying why you want to be an engineer, (2) learning as much as you can about engineering, and (3) developing a step-by-step guide, or "road map," for you to follow.

We noted that achieving a goal requires you to adopt appropriate attitudes and behaviors. We also discussed the importance of effort, in terms of both time-on-task and mental energy.

Last, we explored the importance of the approach you take to your engineering studies. We saw that success not only means that you study <u>hard</u> but also that you study <u>smart</u>.

Three models were then presented to help you understand what a quality education entails.

(1) The first model lists the attributes all engineering graduates should have as mandated by the Accreditation Board for Engineering and Technology (ABET).

(2) The second model focuses on the qualifications that employers seek when considering candidates for entry-level engineering positions.

(3) The third model stresses the importance of "student involvement" to ensure you get a quality education.

Each of these models identifies the knowledge, skills, personal qualities, and behaviors that you need to develop during your college years. Each model also identifies specific areas against which you can assess yourself. Doing periodic personal assessments will call attention to your strengths and areas for improvement.

The chapter closed by talking about the need to structure your life situation so that it supports your goal of graduating with an engineering degree. The gist of this discussion centered on your ability to balance the demands of your school work with outside demands – jobs, family, friends, and all other sources – so that you reserve adequate time to devote to your studies.

References

1. Covey, Stephen R., *The Seven Habits of Highly Effective People*, Free Press, New York, NY, 2004.

2. Tinto, Vincent, *Leaving College: Rethinking the Causes and Cures of Student Attrition, Second Edition*, The University of Chicago Press, 1993.

3. Iacocca, Lee, *Iacocca: An Autobiography*, Bantam Books, New York, NY, 1986.

4. Hansen, James R., *First Man: The Life of Neil A. Armstrong*, Simon & Schuster, New York, NY, 2005.

5. Hickam Jr., Homer H., *Rocket Boys*, Delta Publishing, 2000.

6. Beyer, Kurt W., *Grace Hopper and the Invention of the Information Age,* MIT Press, 2009.

7. Neufeld, Michael, *Von Braun: Dreamer of Space, Engineer of War*, Vintage, 2008.

8. Holtz, Lou, *Wins, Losses, and Lessons: An Autobiography*, William Morrow, New York, NY, 2006.

9. Jacobs, Joseph J., *The Anatomy of an Entrepreneur*, ICS Press, Institute for Contemporary Studies, San Francisco, CA, 1991.

10. Dweck, Carol S., *Mindset: The New Psychology of Success*, Random House, New York, NY, 2006.

11. Accreditation Board for Engineering and Technology (ABET), 111 Market Place, Suite 1050, Baltimore, MD 21202 (*Engineering Criteria 2000* available on ABET webpage: <u>www.abet.org</u>).

12. Hundley, Stephen; Brown, Lynn; and Jacobs, Alan, "Attributes of a Global Engineer: Field-Informed Perspectives, Recommendations, and Implications," Presented at American Society for Engineering Education 2012 Annual Conference, San Antonio, TX, June, 2012.

13. Astin, Alexander W., "Involvement: The Cornerstone of Excellence," *Change*, July/August 1985.

14. Mulinazzi, Tom, "The 60-Hour Rule," *Success 101*, Issue 1, Spring, 1996. (Available from rlandis@calstatela.edu).

PROBLEMS

1. Have any of your teachers or professors ever done anything to make you feel as though you couldn't make it? What did they say or do? Why do you think they said or did that?

2. Discuss the relationship between *success* and *happiness*. What is the definition of each of these words? Does success bring happiness? Can people be happy if they are not successful?

3. Do you have a personal goal of graduating with your bachelor of science degree in engineering? How important is that goal to you? How can you make it more important?

4. Develop a list of 20 goals you would like to accomplish in your lifetime. Be bold!

5. Establish a goal for the grade you want to achieve in each of your courses this term. What GPA would this give you? How would it compare to your overall GPA?

6. List ten benefits that will come to you when you graduate in engineering. Rank them in order of importance to you.

7. List ten tasks that an engineer might perform (e.g., write a report, conduct a meeting, perform a calculation). Rank them in the order that you would most enjoy doing. Explain your ranking.

8. Read a biography of a famous engineer. Write a critique of the biography. Include a discussion of what you learned from the book that will help you succeed in engineering study.

9. Do you believe the statement, "You learn more from your failures than you do from your successes"? Have you ever experienced a significant failure? What was it? What did you learn from that experience?

10. Have you ever achieved anything that others thought you couldn't achieve through sheer determination? What was it?

11. How many hours do you think you should study for each hour of class time in your mathematics, science, and engineering courses? Is this the same for all courses? If not, list four factors that determine how much you need to study for a specific class.

12. Ask one of your professors why he or she chose teaching as a career (rather than professional practice).

13. Would you rather tackle an easy problem or a difficult one? Which do you think would benefit you more? Make an analogy with the task of developing your physical strength.

14. List five things you could do to study "smart" that you are not currently doing. Pick the two most important ones and try to implement them. Prepare a brief presentation for your *Introduction to Engineering* class that discusses your success or lack of success in implementing them.

15. List six things that your professors can do for you beyond classroom instruction.

16. If you spend 100 hours studying, how many of those hours would you be studying alone? How many would you be studying with at least one other student? If you study primarily alone, why? List three benefits of studying with other students.

17. Check off any of the statements below that describe your attitude.

ATTITUDE	
My commitment to success in engineering study is weak.	
I lack confidence in my ability to succeed in engineering study.	
I have a tendency to sabotage my success.	
I tend to blame others for my failures.	
I don't see any need to change myself or to grow or develop.	
I am generally unwilling to seek help from others.	
I tend to procrastinate, putting off the things I need to do.	
I tend to avoid doing things that I don't enjoy.	
I avoid contact with my professors outside of class.	
I prefer to study alone rather than with other students.	

For any of the items you checked, answer the following questions:

a. Is this attitude working for you (positive attitude) or working against you (negative attitude)?

b. If the attitude is working against you, can you change it? How?

18. Rank ABET's list of attributes of engineering graduates presented in Section 1.4 in your order of importance. Meet with your engineering advisor or an engineering professor to discuss your ranking.

19. List ten skills or attributes that you need to work effectively with other people. How can you go about acquiring these skills and attributes?

20. Find out if your engineering college has a list of attributes it strives to impart to its graduates. If so, how does it compare with the list in Section 1.4?

21. Rate yourself on a scale of zero to ten (ten being highest) on the following items:

DESCRIPTION	RATING
Writing skills	
Oral communication skills	
Ability to work on teams	
Commitment to becoming an engineer	
Understanding of professional and ethical responsibility	
Knowledge of contemporary issues	
Recognition of the need for lifelong learning	
Knowledge of contemporary issues	
Computer skills	
Ability to apply knowledge of mathematics	

Ability to apply knowledge of science	
Participation in student organizations	
Studying collaboratively with other students	
Time and energy devoted to studying	
Time spent on campus	
Interaction with faculty members	
Overall grade point average	

22. Rate the items in Problem #21 above on a scale of zero to ten (ten being highest) as to their importance.

23. Develop a method for determining which of the items in Problem #21 need your greatest attention. (Hint: Use the 2x2 matrix below.) Which quadrant contains items that need your greatest attention? Which quadrant contains items that need the least attention?

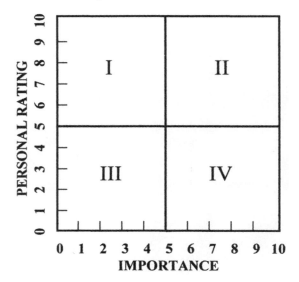

24. From the list in Problem #21, pick the three items that need your greatest attention and the three items that need your least attention. Develop a plan for self-improvement for those that need your greatest attention. Implement the plan.

25. Which of the items in Problem #21 have to do with your skills? With your attitude? With your approach to your studies?

26. Make a list of factors that interfere with your ability to perform academically up to your full potential. How many of these are external (e.g., job, family, friends)? How many are internal (e.g., lack of motivation, poor study habits, etc.)? Which of these interferences can you reduce or eliminate? Develop a plan to do so.

27. Apply the "60-Hour Rule" presented in Section 1.5 to your situation. Based on that rule, how many credit hours should you be taking? How many are you taking? Are you over-committed? What can you do about it?

28. Who are your best friends? Are they engineering majors? How many engineering majors do you know by name? What percentage of the students in your key math, science, and engineering classes do you know? How could you get to know more of them?

CHAPTER 2
The Engineering Profession

Scientists investigate that which already is;
Engineers create that which never has been.

— Theodore Von Karman

INTRODUCTION

This chapter will introduce you to the engineering profession. Look at it as a discussion of "everything you ever wanted to know about engineering" – and then some. Hopefully, when you are finished reading the chapter, you will have a comprehensive understanding of the engineering profession and perhaps have found the engineering niche that attracts you most. This information, coupled with knowledge of the personal benefits you will reap from the profession, is intended to strengthen your commitment to completing your engineering degree.

First, we'll answer the question, "What is engineering?" Through several standard definitions, you'll learn that engineering is essentially the application of mathematics and science to develop useful products or processes. We'll then discuss the engineering design process, which we will demonstrate through a case study of an actual student design project.

Next, we will discuss the rewards and opportunities that will come to you when you earn your B.S. degree in engineering. Having a clear picture of the many payoffs will be a key factor in motivating you to make the personal choices and put forth the effort required to succeed in such a challenging and demanding field of study.

The remainder of the chapter will be devoted to an in-depth look at engineering – past, present, and future.

To look at the past role of engineering in improving the quality of our lives, we will take stock of the Greatest Engineering Achievements of the 20th Century, selected by the National Academy of Engineering.

To look at the present state of engineering, we will examine the various engineering disciplines, the job functions performed by engineers, and the major industry sectors that employ engineers. The North American Industry Classification System (NAICS) will serve as a window into the vast organization of U.S. industry.

To gain some insights about the future of engineering, we will look at the Grand Challenges for Engineering presented by the National Academy of Engineering in 2008. These challenges provide an indication of those fields showing the greatest promise for future growth.

The last section of the chapter will focus on engineering as a profession, including the role of professional societies and the importance of professional registration.

2.1 WHAT IS ENGINEERING?

I'm sure you have been asked, *"What is engineering?"* I remember my grandmother asking me that question when I was in college. At the time, I didn't have much of an answer. Yet, when you think about it, it is a fundamental question, especially for a new engineering student like yourself. So, just what is engineering?

A good starting point for answering this question is the theme of *National Engineers Week*, held each February in honor of George Washington, considered our nation's first engineer. That theme depicts engineering according to its function:

Engineers turn dreams into reality.

Over the years, many variations of this theme have been put forth, from that of the famous scientist Count Rumford over 200 years ago:

> *Engineering is the application of science to the common purpose of life.*

to the current standard definition of engineering provided by the Accreditation Board for Engineering and Technology (ABET):

> *Engineering is the profession in which a knowledge of the mathematical and natural sciences, gained by study, experience, and practice, is applied with judgment to develop ways to utilize, economically, the materials and forces of nature for the benefit of [hu]mankind.*

Harry T. Roman, a well-known New Jersey inventor and electrical engineer, compiled 21 notable definitions of engineering. These are listed in Appendix B.

> **EXERCISE**
>
> Study the 21 definitions of engineering in Appendix B. Then compose your own definition. Write it down and commit it to memory. This may seem like an unnecessary exercise, but I assure you it isn't. Aside from impressing others with a quick informed answer to the question "What is engineering?" this exercise will help clarify your personal understanding of the field.

LEARNING MORE ABOUT ENGINEERING

As you learn more about the field of engineering, you will find there is no simple answer to the question "What is engineering?" Because engineers do so many different things and perform so many different functions, learning about engineering is a lifelong endeavor. Still, there is a variety of ways to start learning about and understanding engineering, one being to tap the tremendous amount of information available online.

One helpful website you should check out is www.eweek.org. At that website you can learn much about both engineering and National Engineers Week. The following additional websites will help further your understanding of engineering:

www.engineeringdegrees101.com
www.futuresinengineering.com
www.discoverengineering.org
www.tryengineering.org
www.careercornerstone.org
www.egfi-k12.org
www.dedicatedengineers.org
www.bls.gov/ooh/architecture-and-engineering/home.htm

Remember, by increasing your knowledge about engineering, you will strengthen your commitment to your studies and your desire to succeed.

ONE LAST POINT. The question is often asked: How is engineering different from science? An excellent answer was provided by astronaut Neil Armstrong in the foreword of *A Century of Innovation: Twenty Engineering Achievements That Changed Our Lives* [1]:

Engineering is often associated with science and understandably so. Both make extensive use of mathematics, and engineering requires a solid scientific basis. Yet as any scientist or engineer will tell you, they are quite different. Science is a quest for "truth for its own sake," for an ever more exact understanding of the natural world. It explains the change in the viscosity of a liquid as its temperature is varied, the release of heat when water vapor condenses, and the reproductive process of plants. It determines the speed of light. Engineering turns those explanations and understandings into new or improved machines, technologies, and processes – to bring reality to ideas and to provide solutions to societal needs.

2.2 THE ENGINEERING DESIGN PROCESS

At the heart of engineering is the *engineering design process*. The engineering design process is a step-by-step method to produce a device, structure, or system that satisfies a need.

Sometimes this need comes from an external source. For example, the U.S. Air Force might need a missile system to launch a 1,000-pound communications satellite into synchronous orbit around the earth.

Other times, the need arises from ideas generated within a company. For example, consumers did not initiate the need for various sizes of little rectangular yellow papers that would stick onto almost anything yet be removed easily when 3M invented "Post-its" [2].

Whatever the source, the need is generally translated into a set of specifications ("specs"). These include performance specifications (e.g., weight, size, speed, safety, reliability), economic specifications (e.g., cost), and scheduling specifications (e.g., production and delivery dates).

YOUR ALARM CLOCK IS AN EXAMPLE

Virtually everything around you was designed by engineers to meet certain specifications. Take the start of your day, for example. You likely wake up to an electrically-powered alarm clock. Every design feature of the clock was carefully considered to meet detailed specifications. The alarm was designed to be loud enough to wake you up but not so loud as to startle you. It may even have a feature in which the sound level starts very low and increases progressively until you wake up. The digital display on your clock was designed to be visible day and night. Batteries may be included to provide redundant power so the alarm will work even if there is a power outage. These batteries must meet life, safety, and reliability requirements. Economic considerations dictated material selection and manufacturing processes. The clock also had to look aesthetically pleasing to attract customers, while maintaining its structural integrity under impact loading, such as falling off your night stand.

STEPS IN THE ENGINEERING DESIGN PROCESS

Now that you have been introduced to the first two steps – identifying the need and then drawing up specifications to meet that need – a complete eight-step engineering design process can be illustrated by the schematic below.

The Engineering Design Process

From this schematic you can see that each step of the design process reflects a very logical, thorough problem-solving process. The customer need or business opportunity (Step 1) leads to a problem definition, including a description of the design specifications (Step 2).

Early in the design process, a number of constraints may be identified. Whatever these constraints may be – e.g., availability of parts and materials, personnel, and/or facilities – the final design must not only meet all design specifications but also satisfy any constraints.

The problem definition, specifications, and constraints will need to be supplemented by additional data and information (Step 3) before the development of possible solutions can begin. This step might, for example, involve learning about new technologies and where information is lacking, research may need to be done.

The process of developing and evaluating possible designs (Steps 4 and 5) involves not only creativity but also the use of computer-aided drafting (CAD), stress analysis, computer modeling, material science, and manufacturing processes. Engineers also bring common sense and experience to the design process. At the conclusion of Step 5, based on a comparative evaluation, the optimal design will be selected.

Step 6 involves implementing the optimal design, which in many cases involves fabrication of a device. Fabrication of several designs may be required in order to test how well each meets the performance specifications. In Step 7, the final design is tested and evaluated, and if necessary, redesigned and retested (Step 8).

Many iterations through the engineering design process may be required before a design is found that meets the need or opportunity and all specifications and satisfies all constraints.

It should be noted that the engineering design process is part of the broader product development cycle that begins with the perception of a market opportunity and ends with the production, sale, and delivery of a product. An excellent resource on this subject is the book *Product Design and Development* by Karl Ulrich and Steven Eppinger [3].

The engineering design process is succinctly stated through the motto of Lamar University's College of Engineering:

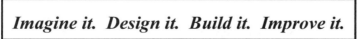

Imagine it. Design it. Build it. Improve it.

HOW THINGS WORK - *REVERSE ENGINEERING*

As an engineering student, you should take every opportunity to learn as much as you can about how things work. Much of your engineering coursework will deal with mathematical modeling of physical problems (analysis). You will get some exposure to the "real world" in your laboratory courses and in courses that focus on engineering design. But as they say, "You can't get enough of a good thing." The more you know and understand the physical world, the better engineer you will be.

I suggest you take initiative in this regard. It's one thing to study about a subject; it's another to take things apart and analyze how they work.

STUDYING ABOUT HOW THINGS WORK. A great source of information on how things work is the "How Stuff Works" website: www.howstuffworks.com. To get a feel for what's there, select something you know little or nothing about but would like to know more about (e.g., da Vinci surgical system, synthesizers, fuel injection systems, optical mice, intelligent highways). And go to the HowStuffWorks website and enter the device name.

You can also learn about how things work and keep up with changing technologies by reading trade magazines such as *Popular Mechanics*, *PC World*, *Popular Science*, *Wired*, and *Discover Magazine*.

Professional engineering societies also have magazines and websites that are good sources of technical information, although some may only be available to members. Examples of those non-members can access are:

IEEE Spectrum – www.spectrum.ieee.org

ASME Mechanical Engineering – www.asme.org/kb/newsletters

ASCE Civil Engineering – www.asce.org/cemagazine

REVERSE ENGINEERING. A more formal topic related to understanding how things work is called *reverse engineering*. Reverse engineering is the process of taking apart a device, object, or system to see how it works in order to duplicate or enhance it. Reverse engineering had its origins in the analysis of hardware for commercial or military advantage. This practice is now frequently used on computer software.

Reverse engineering of hardware might be done out of curiosity or as an academic learning experience. Have some fun! Look for opportunities to take things apart and figure out how they work.

More often reverse engineering is done by businesses to make a three-dimensional record of their products or to assess competitors' products. It's also done to retrieve the source code of a software program, improve the performance of a program, correct an error in the program, identify a virus, or adapt a program written for one microprocessor for use with another.

REFLECTION

Before you read the next section of this chapter, reflect on the task of designing and building a human-powered helicopter. Do you think this is possible? Do you think it has ever been done? How long could a human-powered helicopter stay aloft? What altitude could it reach? Make a sketch of how you think a human-powered helicopter would look.

2.3 CASE STUDY: HUMAN-POWERED HELICOPTER

The eight steps of the engineering design process make most sense when they are seen in action. The following case study of the design, construction, and test of a human-powered helicopter by a team of faculty and students at the University of Maryland will enable you to see each step of the process at work.

STEP 1 - CUSTOMER NEED OR BUSINESS OPPORTUNITY

In 2008 a team of University of Maryland faculty and students was formed to try to win the American Helicopter Society's Igor I. Sikorsky Human-Powered Helicopter Prize.

The Sikorsky Prize was initially established in 1980 to promote fulfillment of the dream of human-powered hovering flight. A $250,000 cash award was pledged to the first team to fly a helicopter, under human power only, for at least 60 seconds, and reach an altitude of three meters above the ground momentarily during the flight without drifting outside of a ten-square-meter area.

In this case, the opportunity or need – the first step in the engineering design process – was created by the American Helicopter Society. (Information about the Sikorsky Prize can be found at: www.vtol.org/awards-and-contests/human-powered-helicopter.)

The idea of a human-powered helicopter was inspired more than 500 years ago by famous artist and inventor Leonardo Da Vinci's conceptual illustration of an "airscrew." Da Vinci's flying machine consisted of a 15-ft-diameter airscrew designed to be powered by four men. Although the machine was never built and could not have flown, Da Vinci's airscrew concept is the basis for the propellers used in propeller-driven aircraft and for rotors used in helicopters.

Because of the university's mascot, the Terrapin, the University of Maryland team took on the name *Gamera* from a giant flying turtle in Japanese horror movies.

The Gamera team was well aware that although several teams had attempted to win the Sikorsky Prize in the 32 years since it was established, none had been successful. Several human-powered helicopters had hovered for short times including the Da Vinci III (eight seconds) built by a team from Cal Poly San Luis Obispo in 1989 and the Yuri-I (19.5 seconds) built by a team from Nihon University in Japan in 1994.

Knowing that so few teams had even tried for the Sikorsky Prize and that those who did fell far short of the requirements for the prize made the challenge even more exciting for the Gamera team.

They also knew that a recent paper titled "On the Possibility of Human-Powered Vertical Flight" [4] had concluded that, under the most optimistic assumptions, "hovering . . . is in principle possible with a single crew member, provided that the rotorcraft has enough longitudinal and lateral stability." At least the task was theoretically possible!

STEP 2 - PROBLEM DEFINITION/SPECIFICATIONS AND CONSTRAINTS

The primary design specifications, Step 2 of the engineering design process, were established by the regulations of the Sikorsky Human-Powered Helicopter Competition. They included the following requirements:

- Type of machine: Rotary wing capable of vertical takeoff and landing
- Vehicle size: No limitation
- Flight requirements: Hover for 60 seconds
- Altitude of flight: Exceed three meters momentarily
- Drift: A reference point on the non-rotating part of the machine stays within the confines of 10 x 10 meter square.
- Crew: No limitation on number. One member of crew shall be non-rotating.
- Power: Powered by crew during entire flight including accelerating rotor up to takeoff speed
- Control: Controlled by crew (No remote control)
- Energy storage devices: None permitted
- Lighter-than-air gases: Prohibited
- Jettison: No part of the machine shall be jettisoned nor any member of the crew leave the aircraft during flight.
- Ground conditions: Level ground (<1/100 slope)
- Air conditions: Still air (< 1 meter/sec)

For the Gamera team, these requirements led to additional problem definitions and specifications. Who would lead the team? What additional capabilities would need to be represented on the team? How much money would the project cost? How would it be financed? What facilities would be required? What would be the general configuration of the machine? What materials would be needed to fabricate it? What sort of manufacturing techniques would be needed? Could the vehicle be tested outdoors or would the capacity of available indoor facilities limit the size of the vehicle?

STEP 3 - DATA AND INFORMATION COLLECTION

The Gamera team realized it faced many daunting challenges. Before developing alternative designs, the team first had to collect extensive data and information – Step 3 of the engineering design process. They needed to learn the technologies associated with low-speed airfoil design, design of lightweight structures, rotor ground effects, vehicle stability, power transmission, and human power capability.

Valuable information could be obtained from reviewing the human-powered helicopter literature and from the experience of the Da Vinci III [5] and Yuri I [6] projects. Lessons could also be learned from the successful human-powered aircraft projects: the Gossamer Condor [7] and the Gossamer Albatross [8].

A major issue was ground effect. Ground effect is a well-known phenomenon in which rotorcraft experience an increase in performance when operating near the ground. Since no data existed for the large rotors and low rotation speeds expected for the Gamera I vehicle, a comprehensive research and test program was needed. The information from these studies would be key since ground effect would reduce the power required to produce a specific amount of lift by as much as 60 percent.

Extensive testing of both sub-scale and full-scale rotors was done to gain information needed to optimize the rotor blade designs. Variables examined included rotational speed, blade pitch, and height above the ground,

Other studies included pilot power production for various lengths of time from ten seconds to 60 seconds both with and without hand cranking.

STEP 4 - DEVELOPMENT OF ALTERNATIVE DESIGNS

Once the team had collected sufficient basic data, it was time to move to Step 4 of the design process: developing alternative designs. In the design of the Gamera vehicle, many design tradeoffs had to be considered and many decisions had to be made. The team knew that the major component parts of a human-powered helicopter are:

- Rotors
- Airframe
- Cockpit
- Power transmission system
- Power plant (pilot)

The team faced many questions and design tradeoffs for each of these components.

ROTORS. For the rotors, many choices had to be made. How many rotors? What airfoil shape would be used for each rotor? What should be the radius of each rotor? The cord length? The angle of attack? Should the rotor blades be tapered and/or twisted? What would be the allowable weight of each rotor, and could that weight be achieved while still maintaining structural integrity? How stiff would each rotor need to be? What tip speed should the rotor operate at?

AIRFRAME. As the support for the rotors and the cockpit, the airframe needed to be as light-weight as possible while still maintaining structural integrity.

COCKPIT. The cockpit needed to be comfortable and structurally sound while being as light-weight as possible. It also needed to be able to accommodate pilots of different heights and weights.

POWER TRANSMISSION SYSTEM. Choices existed for the power transmission system as well. How would the power be transmitted from the pilot to the rotors? Would the power be generated by legs only or could arms be used as well? How should the power be transmitted from the pilot-side pulley to the rotor pulley? Choices might be chain drive, belt drive, shaft drive, or winch drive. What sizes should the pulleys be?

POWER PLANT (PILOT). One of the most important aspects of the project was the selection and training of the pilot. What would be the optimal weight of the pilot? How should the pilot be trained for maximum power output? How much power could be generated for five seconds? For 20 seconds? For 60 seconds? What would be the optimal RPM for the pilot to maximize power input?

Coupled with optimizing the design for each component was the critical issue of how the components would be configured into an overall vehicle design.

STEP 5 - EVALUATION OF DESIGNS/SELECTION OF OPTIMAL DESIGN

This is one of the most difficult, challenging, and time-consuming steps in the engineering design process. For many engineers, however, it is also the most interesting and rewarding one, for here is where ideas really begin to turn into reality.

A brief overview of the design decisions made by the Gamera team is presented here. You can find detailed discussion of the Gamera I design features in "Design and Development of Gamera: A Human Powered Helicopter from the University of Maryland" [9].

The first fundamental design decision made was that the vehicle would be a quadrotor helicopter with an airframe consisting of interconnecting trusses and a cockpit. As indicated by the diagram, the Gamera I design consisted of an X-shaped fuselage frame spanning 63 ft. At the terminus of each end of the frame resides a 42.6-ft-long rotor.

Overall vehicle design weight was selected to be 101 pounds broken down by components as follows:

Rotor blades – 55%
Airframe – 30%
Transmission – 6%
Cockpit – 9%

The airframe design consisted of four 31.4-ft lightweight trusses. The allowable design weight of eight pounds for each truss was to be achieved using unidirectional carbon fiber tubes. Significant analysis and testing of one-third scale models of the truss configuration were conducted to ensure adequate stiffness and resistance to buckling.

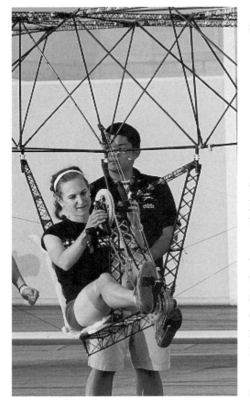

The four rotors were designed to the following specifications:

Airfoil: Eppler387
Rotor radius: 21.3 ft.
Chord: 3.3 ft.
Taper: None
Weight (Eight rotor blades): 58.3 lbs.
Design speed: 17-18 RPM

Design weight for the cockpit was set at 9.5 lbs. The cockpit design consisted of three stiff, 2-D trusses (see photo) connecting the seat, hand cranks, and foot cranks to the airframe structure at three nodes.

Power from the pilot would be transferred to the rotors via hand and foot pedals in the cockpit suspended beneath the aircraft structure. A string drive system, similar to a rod and reel, was chosen based on low weight and high efficiency. A high strength fiber line wound around each rotor pulley was "reeled in" to the pilot pulley through the foot cranks and hand cranks.

Through tradeoff studies, the pilot design weight was selected to be 107 pounds. Testing indicated that a pilot of that weight could generate 0.79 hp (590 watts) for a 10-second flight and 0.76 hp (565 watts) for a 20-second flight.

STEP 6 - IMPLEMENTATION OF OPTIMAL DESIGN

Once the design phase was completed there was no time to waste. The Gamera team was well aware that other teams were chasing the Sikorsky Prize, including the formidable AeroVelo team from Canada.

The fabrication and construction of a 101-pound vehicle that would fill a gymnasium brought significant challenges. Because of the size of the vehicle, components had to be modular and easily assembled on site. Due to the fragile nature of each component, backup parts were needed.

Particular care needed to be taken in constructing the eight 7.4-pound rotors because they represented a significant portion (55 percent) of the vehicle weight. Building the 10.6-ft-long rotors with spars made of trusses of carbon fiber composite material, polystyrene foam, and balsa strips, and covered by a lightweight Mylar film took significant time and handiwork.

The four trusses comprising the airframe were made from "microtrusses" in which spindles of carbon fiber were wrapped around a truss of three carbon-fiber rods. Fabrication of the airframe required intricate handwork, which introduced the potential for human error, so great care was required to prevent unexpected failures.

Finally, the cockpit and power transmission system were constructed. The cockpit consisted of stiff, 2-D planar trusses and foot pedals and cranks for delivering the power to the rotors. The rotor side pulleys were made of an expanded polystyrene foam core with four poles reinforced with carbon composite rods (see photo on the right).

STEP 7 - TEST AND EVALUATION OF GAMERA I

On May 11, 2011, almost three years in the making, the Gamera I vehicle was finally ready for testing. The pilot was 24-year-old University of Maryland life sciences graduate student Judy Wexler. She would have to generate sustained power approaching 0.75 hp.

First, three test runs were made in which the rotors were brought up to the design speed of 18 RPM without liftoff. Finally, it was time to attempt liftoff: The aircraft became airborne a few inches above the ground for at least four seconds, and the flight was the first ever by a woman.

A few weeks later, on July 13, the Gamera I (again piloted by Wexler) flew in controlled hover for 11.4 seconds. The attempt set a new United States record for flight duration. The team was awarded the Igor I. Sikorsky International Trophy from American Helicopter Society. Successful Gamera I flights can be viewed at www.YouTube.com/watch?v=n-qFhcL9mg4.

The project was deemed a huge success. Records had been set, and Gamera I was the first human-powered helicopter to lift off in more than 17 years. However, it became clear through observations of the vehicle dynamics and pilot fatigue that Gamera I was not capable of achieving the flight conditions required for the Sikorsky Prize, and the vehicle was retired.

STEP 8 - REDESIGN

The Gamera team knew winning the Sikorsky Prize would require improved performance, so they immediately set about re-designing the vehicle toward that end. The Gamera II project was born.

The team would benefit from the many lessons learned in the design, construction, and testing of Gamera I.

The same overall quadrotor layout of Gamera I was retained due to familiarity with the design and the stability it offered. The four rotor diameters were kept at 13 meters (due to the space limitation of indoor testing locations).

Substantial improvements were made in vehicle weight. The rotor weight was reduced from 58 pounds to 35 pounds, and the airframe truss weight was reduced from 32 pounds to 19 pounds using specially developed micro-truss members and improved manufacturing techniques.

The rotor blades were redesigned with a thicker Selig S8037 airfoil, and a 3:1 taper was incorporated. These improvements reduced bending deflections of the rotor blades, which increased the ground effect (and hence reduced the power needed to hover) and reduced the danger of the rotor blades striking the airframe structure overhead.

Pilot recruiting and training were expanded. A fly-wheel was added to smooth the power delivery and structural improvements were made to the cockpit to improve power transfer from the pilot to the rotors.

On June 21, 2012, the Gamera II vehicle piloted by Maryland mechanical engineering graduate student Kyle Gluesenkamp set a world record for flight duration of 49.9 seconds. Several weeks later a record height of eight feet was reached for a shorter time.

However, on September 1, 2012, the Gamera II crashed after momentarily reaching a record altitude of 9.4 feet, just five inches below the 9.84 feet required for the Sikorsky Prize. You can view all of the major Gamera II flights at www.YouTube.com (Search under "Gamera II"). You can read more about the Gamera I and Gamera II projects in a number of excellent papers written by University of Maryland students and faculty [10, 11].

Whether the Gamera team (or any other team) will have won the Sikorsky Prize by the time you read this is anybody's guess. As the team learned, flying a human-powered helicopter for 60 seconds to a momentary height of three meters while staying in a 10 x 10 meter area is no easy task.

THE NEEDS AND OPPORTUNITIES FOR ENGINEERING DESIGN ARE BOUNDLESS

The purpose of chronicling the University of Maryland's human-powered helicopter project was to illustrate the engineering design process in action. Now that you have seen the logic and demand that each step of the process entails, you should easily be able to come up with a list of many other problems, needs, and opportunities that would suit its step-by-step approach. Here are just a few ideas that occurred to me. What ideas would you add to this list? Remember, it is entirely possible that, down the road, you will be the engineer who turns one of these needs into reality:

- A device carried by a police officer that would detect a bullet fired at the officer and intercept it

- A device that would mark the precise location of a football when the referee blows the whistle

- A fail-safe system that prevents an automobile from being stolen

- A device that would program a DVR to skip the commercials while taping your favorite TV show

- A machine that would serve ping-pong balls at different speeds and with different spins

- A device that identifies vehicles that are carrying explosives

- A car alarm that goes off if the driver falls asleep.

- An in-the-ground, AC-powered sound/vibration emitter to repel gophers

- A device that cuts copper tubing in tight places

- An in-home composting and recycling system that eliminates the need for sewer or septic systems

- A device that prevents elderly people from being injured when they fall down

- A portable solar-charged lamp that can provide two hours of reading light

- Software that turns your computer monitor into a mirror

- A travel toothbrush with toothpaste dispenser attached

- A cordless hairdryer that can be used when camping

- An affordable, fuel-cell-powered automobile that only emits water vapor

- A system that continues to tape your favorite morning radio program after you arrive at work so you can listen to it on the way home

- A high-rise building with an "active suspension system" that responds to ground movement (earthquakes)

REFLECTION

Think about "customer needs and business opportunities" for each of the engineering design ideas listed above. Could you get excited about working on any of those items? Could you add to this list by thinking of something that would improve the quality of life that is not currently available?

2.4 REWARDS AND OPPORTUNITIES OF AN ENGINEERING CAREER

Engineering is a unique and highly selective profession. Among the 128 million people employed in the United States, only about 1.5 million (1.2 percent) list engineering as their primary occupation [12]. This means the overwhelming majority of people employed in this country do something **other than engineering**.

These employment figures are reflected by national higher-education statistics. Engineering typically represents less than five percent of college graduates, as the following table shows [13]:

Major	Number of 2009/10 College Graduates	Percent of Total
Business	358,293	21.7%
Social Sciences	172,780	10.5%
Health Professions	129,634	7.9%
Science and Mathematics	125,809	7.6%
Education	101,265	6.1%
Psychology	97,216	5.9%
Visual and Performing Arts	91,802	5.6%
Communication Studies	81,266	4.9%
Engineering	**72,654**	**4.4%**
TOTAL	1,650,014	100.0%

So why choose to study engineering? Why strive to become one of those 4.4 percent of college graduates who receive their B.S. degree in engineering? I'll tell you why. **Job satisfaction!**

JOB SATISFACTION – AN OVERARCHING ISSUE

What would you say is the number one cause of unhappiness among people in the United States? Health problems? Family problems? Financial problems? No. Studies have shown that, by far, the number one cause of unhappiness among people in the U.S. is **job dissatisfaction**. Furthermore, Americans are growing increasingly unhappy with their jobs. A study conducted in 2012 for the Conference Board, a leading business membership and research organization, indicated that only 47.2 percent of all Americans are satisfied with their jobs, down from 61 percent in 1987 [14].

Do you know people who dislike their job? People who get up every morning and wish they didn't have to go to work? People who watch the clock all day and can't wait until their workday is over? People who look forward to Fridays and dread Mondays? People who work only to earn an income so they can enjoy their time off? Maybe you have been in one of these situations. Lots of people are.

Throughout my career, it has been very important to me to enjoy my work. After all, I spend eight hours or more a day, five days a week, 50 weeks a year, for 30 or 40 years working. This represents about 40 percent of my waking time. Which would you prefer? Spending 40 percent of your life in a career (or series of jobs) you despise? Or spending that 40 percent in a career you enjoy and love? I'm sure you can see why it is extremely important to find a life's work that is satisfying, work that you <u>want</u> to do.

Engineering could very well be that life's work. It certainly has been for me and for many of my colleagues over the years. But what exactly does "job satisfaction" mean?

What is it about engineering that is so satisfying? My personal "Top Ten List" of the personal and career rewards of an engineering career is presented below. Although your list may differ from mine, I am going to discuss each briefly – if only to help you realize more fully the many rewards, benefits, and opportunities an engineering career holds for you.

Ray's Top Ten List

1. **Varied Opportunities**
2. **Challenging Work**
3. **Intellectual Development**
4. **Social Impact**
5. **Financial Security**
6. **Prestige**
7. **Professional Environment**
8. **Understanding How Things Work**
9. **Creative Thinking**
10. **Self-Esteem**

After studying my list and developing your own, hopefully you will find yourself more determined to complete your engineering studies. You may also find yourself somewhat puzzled by the skewed statistics that opened this section. With so many benefits and job opportunities a career in engineering promises, you'd think that college students would be declaring engineering majors in droves.

I guess engineering really is a unique and highly selective profession. Consider yourself lucky to be one of the "chosen few."

1. Varied Opportunities

While the major purpose of this chapter is to help you understand the engineering profession, you have just skimmed the surface thus far. Your introduction to the engineering field has largely been a "functional" one, starting with the idea that engineering is the process of "turning dreams into reality," followed by a detailed look at the engineering design process: more function.

As you'll learn subsequently, engineering entails much more than just functions governed by a rigid eight-step design process. I like to think of engineering as a field that touches almost every aspect of a person's life. I often point out to students that the day you walk up the aisle to receive your degree in engineering, you have closed no doors. **There is nothing you cannot become from that time forward!** Doctor. Lawyer. Politician. Teacher. Astronaut. Entrepreneur. Manager. Salesperson. Practicing engineer. All these and many others career opportunities are possible.

Here are examples of people educated as engineers and the professions they ended up in:

ENGINEER	PROFESSION
Neil Armstrong	Astronaut/First Person on Moon
Herbert Hoover	President of the United States
Jimmy Carter	President of the United States
Alfred Hitchcock	Film Director/Producer
Eleanor Baum	First Woman Dean of Engineering
Herbie Hancock	Jazz Musician
Frank Capra	American Film Director
Paul MacCready	Inventor/Winner of Kremer Prize
Ellen Ochoa	Space Shuttle Astronaut
Hyman G. Rickover	Father of the Nuclear Navy
Bill Nye	Host of TV Show "Bill Nye, The Science Guy"
Boris Yeltsin	President of Russia
Alexander Calder	Sculptor
Bill Koch	Yachtsman/Captain of America Cup Team
W. Edwards Deming	Father of Modern Management Practice (TQM)
Grace Murray Hopper	U.S. Navy Rear Admiral/Computer Engineer
Ming Tsai	Restaurateur and Star on TV's Food Network

Hu Jintao	President of the People's Republic of China
Montel Williams	Syndicated Talk Show Host
John H. Sununu	Political Pundit/Governor of New Hampshire
Samuel Bodman	U.S. Secretary of Energy
Donald Thompson	CEO and President, McDonald's Corp.
Rowan Atkinson	British Actor/Comedian/Screenwriter
Rudolph Diesel	Inventor of the Diesel Engine
Michael Bloomberg	Billionaire/Mayor of New York City
Lonnie G. Johnson	Inventor (SuperSoaker®)
A. Scott Crossfield	X-15 Test Pilot
Don Louis A. Ferre	Governor of Puerto Rico
Yasser Arafat	Palestinian Leader/Nobel Peace Prize Laureate
Tom Landry	Dallas Cowboys' Head Coach
Igor Sikorsky	Inventor of Single Rotor Helicopter
Mohamed Morsi	President of Egypt
Shiela Widnall	Secretary of the Air Force
David A. Wolf	Astronaut/Medical Doctor/Electrical Engineer
Robert A. Moog	Father of Synthetic Music
Chester Carlson	Inventor of Xerox Process
John A. McCone	Director of Central Intelligence Agency
Arthur C. Nielsen	Developer of Nielsen TV Ratings

Although none of the above individuals ended up working as a practicing engineer, I expect they would all tell you that their engineering education was a key factor in their subsequent successes. You can learn more about these and other famous "engineers" from these websites: www.sinc.sunysb.edu/Stu/hnaseer/interest.htm and www.engology.com/engpg5profiles.htm

---

Personal Story

When I was an engineering student, I had no idea that the career path I have taken even existed. After completing my B.S. and M.S. degrees in Mechanical Engineering at MIT, I worked for five years as a practicing engineer at Rocketdyne, a Division of Rockwell International. Through part-time teaching to supplement my salary, I developed an interest in an academic career and was offered a position on the engineering faculty at California State University, Northridge.

Although I enjoyed teaching, my interests shifted more to administration and working with students outside of the classroom. I started the first Minority Engineering Program in California and directed it for ten years. The administrative and management experience I gained led me to the position of Dean of Engineering.

My engineering career thus evolved from practicing engineering to teaching it; from teaching it to creating and directing a special program for minority engineering students; and finally from directing a program to managing an entire engineering college.

Within engineering practice itself there is an enormous diversity of job functions. There are analytical engineers, design engineers, test engineers, development engineers, sales engineers, and field service engineers. The work of analytical engineers most closely resembles the mathematical modeling of physical problems you do in school. But only about ten percent of all engineers fall into this category, pointing to the fact that engineering *study* and engineering *work* can be quite different.

- If you are imaginative and creative, **design engineering** may be for you.

- If you like working in laboratories and conducting experiments, you might consider **test engineering**.

- If you like to organize and expedite projects, look into becoming a **development engineer**.

- If you are persuasive and like working with people, **sales or field service engineering** may be for you.

Later in this chapter, we will examine the wide variety of engineering job functions in more detail. Then, in Chapter 8, we will explore less traditional career paths for which engineering study is excellent preparation, such as medicine, law, and business.

2. CHALLENGING WORK

Do you like intellectual stimulation? Do you enjoy tackling challenging problems? If so, you'll get plenty of both in engineering. Certainly, during your period as an engineering student, you will face many challenging problems. But, as the saying goes, "you ain't seen nothing yet." When you graduate you'll enter the engineering work world, where there is no shortage of challenging, open-ended problems.

By "open-ended," I mean there is rarely one "correct" solution, unlike many of the problems you are assigned in school. Open-ended problems typically generate many possible solutions, all of which equally meet the required specifications. Your job is to select the "best" one of these and then convince others that your choice is indeed the optimal one.

It certainly would be helpful if you had more exposure to open-ended problems in school. But such problems are difficult for professors to create, take more time for students to solve, and are excessively time-consuming to grade. Regardless, however, of the kind of problem you are assigned (open-ended or single answer, in school or the engineering work-world), they all challenge your knowledge, creativity, and problem-solving skills. If such challenges appeal to you, then engineering could be a very rewarding career.

3. INTELLECTUAL DEVELOPMENT

Engineering education exercises your brain much the way weight-lifting or aerobics exercises your body – and the results are remarkably similar. The only difference is that physical exercise improves your body, while mental exercise improves your mind. As your engineering studies progress your abilities to solve problems and think critically will grow stronger.

This connection between mental exercise and growth is by no means "news" to educators. But recent research in the cognitive sciences has uncovered knowledge that explains <u>how</u> and <u>why</u> this process works [15]. We now know, for example, that the brain is made up of as many as 180 billion neuron cells. Each neuron has a very large number of tentacle-like protrusions called dendrites. The dendrites make it possible for each neuron to receive signals (synapses) from thousands of neighboring neurons. The extent of these neural networks is determined in large part by the demands we place on our brains – i.e., the "calisthenics" we require of them. So the next time your find yourself reluctant to do a homework assignment or study for a test, just think of all those neural networks you could be building.

One of the things I value most about my engineering education is that it has developed my logical thinking ability. I have a great deal of confidence in my ability to deal effectively with problems. And this is not limited to engineering problems. I am able to use the critical thinking and problem-solving skills I developed through my engineering education to take on such varied tasks as planning a vacation, searching for a job, dealing with my car breaking down in the desert, organizing a banquet to raise money, purchasing a new home, or writing this book. I'm sure you also will come to value the role your engineering education plays in your intellectual growth.

4. SOCIAL IMPACT

I hope you are motivated by a need to do something worthwhile in your career: something to benefit society. Engineering can certainly be an excellent career choice to fulfill such humanitarian goals.

The truth is, just about everything engineers do benefits society in some way. Engineers develop transportation systems that help people and products move about so easily. Engineers design the buildings we live and work in. Engineers devise the systems that deliver our water and electricity, design the machinery that produces our food, and develop the medical equipment that keeps us healthy. Almost everything we use was made possible by engineers.

Depending on your value system, you may not view <u>all</u> engineering work as benefiting people. Some engineers, for example, design military equipment like missiles, tanks, bombs, artillery, and fighter airplanes. Others are involved in the production of pesticides, cigarettes, liquor, fluorocarbons, and asbestos. As an engineer, you will need to weigh the merits of such engineering functions and make your career choices accordingly.

My view is that engineering holds many more beneficial outcomes for society than detrimental ones. For example, opportunities exist for engineers to use their expertise in projects designed to clean up the environment, develop prosthetic aids, develop clean and efficient transportation systems, find new sources of energy, solve the world's hunger problems, and improve the standard of living in underdeveloped countries.

5. FINANCIAL SECURITY

When I ask a class of students to list the rewards and opportunities that success in engineering study will bring them, money is almost always number one. In my "Top Ten List," it's number five. It's not that engineers don't make good money. They do! It's just that money is not a primary motivator of mine.

I've always held the view that if you choose something you like doing, work hard at it, and do it well, the money will take care of itself. In my case, it has. Of course, you may discount my philosophy because of my credentials and career successes. But remember, my engineering career began much the same way yours will: working in industry as a practicing engineer. My subsequent career moves, however, were never motivated by money alone.

I hope you too don't make money your primary reason for becoming an engineer. Other reasons, like challenging work, intellectual development, and opportunities to benefit society hopefully will prove to be more important factors. If they are, you will find the quality of your life enriched tremendously. And I guarantee the money will take care of itself, as it has for me.

Let's not lose sight of reality, however. If you do become an engineer, **you will be rewarded financially**. Engineers, even in entry-level positions, are well paid. In fact, engineering graduates receive the highest starting salary of any discipline, as shown in the data below for 2012 graduates [16].

Beginning Offers to 2012 Bachelor's Degree Graduates

Discipline	Avg. Salary
Engineering	**$60,639**
Computer Sciences	60,038
Business	51,541
Health Sciences (including Nursing)	46,567
Mathematics and Sciences	42,355
Communications	42,286
Education	39,080
Humanities & Social Sciences	36,319
Average for All Disciplines	**$44,259**

As indicated, the $60,639 average starting salary for 2012 engineering graduates is 37 percent higher than the $44,259 average starting salary for all bachelor's degree graduates.

If the starting salary data has not convinced you that engineering is a financially rewarding career, perhaps you will be convinced by the fact that many of the world's wealthiest people started their careers with a degree in engineering. You will find a listing of some of these people in Appendix C. As

reported by *Forbes Magazine* [17], the personal wealth of these individuals ranges from a high of $69 billion down to $5.3 billion. I hope this brings the idea home that "the sky is the limit."

EXERCISE

Pick one of the 30 individuals listed in Appendix C among the world's wealthiest engineers. Find out as much as you can about the person by googling his or her name. What was their engineering discipline in college? What did they do early in their career? How did they become so wealthy? What lessons can you learn from their success?

6. PRESTIGE

What is *prestige*? The dictionary defines it as "the power to command admiration or esteem," usually derived from one's social status, achievements, or profession. Engineering, both as a field of study and a profession, confers prestige. You may have already experienced the prestige associated with being an engineering major. Perhaps you have stopped on campus to talk with another student and during the conversation, he or she asked, "What's your major?" What reaction did you get when you said, "Engineering"? Probably one of respect, awe, or even envy. To non-engineering majors, engineering students are "the really smart, studious ones." Then, if you reciprocated by asking about that student's major, you may wish you hadn't after getting an apologetic response like, "I'm still undecided."

This hypothetical conversation between an engineering and non-engineering student is not far-fetched. In fact, variations of it take place all the time. Everyone knows that engineering study requires hard work, so people assume you must be a serious, highly capable student.

I often ask students to name a profession that is more prestigious than engineering. Medicine always comes up first. I tend to agree. Physicians are well paid and highly respected for their knowledge and commitment to helping people live long and healthy lives. So if you think you want to be a medical doctor and have the ability, arrange to meet with a pre-med advisor as soon as possible and get started on your program. I certainly want to have the most capable people as my doctors.

After medicine, law and accounting are typically cited as more prestigious professions than engineering. Here, however, I disagree, arguing against these and **every other profession** as conferring more prestige than engineering. Anyone who knows anything about engineering would agree that engineers play critical, ubiquitous roles in sustaining our country's international competitiveness, in maintaining our standard of living, in ensuring a strong national security, in improving our health, and in protecting public safety. I can't think of any other profession that affects our lives in so many vital, significant ways.

Engineers are critical to our:

International competitiveness

Standard of living

National security

Personal health

Public safety

7. PROFESSIONAL WORK ENVIRONMENT

Although engineers can perform a variety of functions and work in many different settings, most new engineering graduates are hired into entry-level positions in "hi-tech" companies. While the nature of your work and status within the company may quickly change, there are certain standard characteristics of all professional engineering work environments.

For one, you will be treated with respect – both by your engineering colleagues and by other professionals. With this respect will come a certain amount of freedom in choosing your work and, increasingly, you will be in a position to influence the directions taken by your organization.

As a professional, you also will be provided with adequate workspace, along with whatever equipment and staff support you need to get your work done.

Another feature of the engineering work environment is the number of opportunities you will have to enhance your knowledge, skills, self-confidence, and overall ethos as a professional engineer. Experienced engineers and managers know that new engineering graduates need help in making the transition from college to the "real world." From the outset, then, your immediate supervisor will closely mentor you, giving you the time and guidance to make you feel at home in your new environment. Your supervisor will carefully oversee your work assignments, giving you progressively more challenging tasks and teaming you with experienced engineers who will teach you about engineering and the corporate world.

Once you are acclimated to your position, your company will see to it that your engineering education and professional development continue. You will frequently be sent to seminars and short courses on a variety of topics, from new engineering methods to interpersonal communications. You may be given a travel allotment so you can attend regional or national meetings of professional engineering societies. You also may discover that your company has a reimbursement program to pay your tuition and fees for courses at a local university for professional development or graduate education.

You can expect yearly formal assessments of your performance, judged on the merits of your contributions to the company. As a professional, you will not be required to punch a clock, for your superiors will be more concerned about the quality of your work, not your "time-on-tasks." If you have performed well, you can usually expect a salary increase, plus other bonuses for a job particularly well done. Promotions to higher positions are another possibility, although they generally have to be earned over an extended period of time.

Finally, as a professional, you will receive liberal benefits, which typically include a retirement plan, life insurance, medical and dental insurance, sick leave, paid vacations and holidays, and savings or profit-sharing plans.

8. UNDERSTANDING HOW THINGS WORK

Do you know why golf balls have dimples on them? Do you understand how the loads are transmitted to the supports on a suspension bridge? Do you know what nanotechnology is? How optical storage devices work? How fuel cells work? When you drive on a mountain road, do you understand why the guard rails are designed the way they are? Do you know why split-level houses experience more damage in earthquakes? Do you know why we use alternating current (AC) rather than direct current (DC)? One of the most valuable outcomes of my engineering education is understanding how things around me work.

Furthermore, there are many issues facing our society that depend on an understanding of technology. Why are there so few zero-emission electric vehicles rather than cars powered by highly polluting internal combustion engines? Should we stop building nuclear reactors? Will we be ready to tap alternate energy sources when the earth's supply of oil becomes prohibitively costly or runs out?

Should we have supersonic aircraft, high-speed trains, and automated highways? Is it technically feasible to develop a "Star Wars" defense system that will protect us against nuclear attacks? Why are the Japanese building higher quality automobiles than we are? Can we produce enough food to eliminate world hunger? Do high-voltage power lines cause cancer in people who live near them?

Your engineering education will equip you to understand the world around you and to develop informed views regarding important social, political, and economic issues facing our nation and the world. Who knows? Maybe this understanding will lead you into politics.

9. CREATIVE THINKING

Engineering is by its very nature a creative profession. The word "engineer" comes from the same Latin word *ingenium* as the words "genius" and "ingenious." This etymological connection is no accident: Engineers have limitless opportunities to be ingenious, inventive, and creative. In the next section, we will talk about engineering in the past by listing the "Greatest Engineering Achievements of the 20th Century." You can be sure that creativity played a major role in each of these achievements.

Sometimes new engineering students have difficulty linking "creativity" with "engineering." That's because, at first glance, the terms are likely to invoke their stereotypical connections: "creativity" with art; "engineering" with math, science, and problem solving. The truth, though, is that creativity is an essential ingredient of engineering. Consider, for example, the following definition of "creativity" from *Creative Problem Solving and Engineering Design* [18]:

> *Playing with imagination and possibilities while interacting with ideas, people, and the environment, thus leading to new and meaningful connections and outcomes*

This is just what engineers do. In fact, this definition of "creativity" could almost be a definition of "engineering."

To experienced engineers who regularly engage in solving open-ended, real-world problems, the need for creativity in the engineering process is a given. Creativity is particularly important, for example, during Steps 4, 5, and 6 of the engineering design process described in Section 2.2, which involve developing and evaluating alternative solutions, followed selecting the best one. Without an injection of creativity in these steps, the actual best solution may be overlooked entirely.

However, these are not the only steps of the engineering design process that involve creativity. Indeed, creativity enters into <u>every</u> step of the process. It would be a good exercise for you to review the eight steps of the engineering design process to see how creativity can come into play at each step.

Beyond the engineering process itself, the need for engineers to think creatively is greater now than ever before, because we are in a time when the rate of social and technological changes has greatly accelerated. Only through creativity can we cope with and adapt to these changes. If you like to question, explore, invent, discover, and create, then engineering would be an ideal profession for you.

> *A wonderful place to explore the way human creativity in art, technology, and ideas has shaped our culture is The Engines of Our Ingenuity web page:* <u>www.uh.edu/engines</u>.
>
> *There you will find the text of more than 2,800 episodes originally presented on National Public Radio for more than two decades by John Lienhard, professor of mechanical engineering at the University of Houston.*

10. SELF-ESTEEM

Self-esteem is a critically important factor in virtually every aspect of our life. It influences what we choose to do, how we treat others, and whether we are happy or not. In Chapter 6 we will discuss self-esteem as a fundamental human need.

As you will learn, self-esteem is made up of two interrelated components:

Self-efficacy - your sense of competence

Self-respect - your sense of personal worth

Through your career, you will have a unique opportunity to enhance your self-esteem by building your self-efficacy <u>and</u> your self-respect.

As you gain experience, you will become more knowledgeable and more technically competent. Your ability to think both critically and creatively will improve. Your communication skills will strengthen, as will your effectiveness in working on teams.

Additionally, your understanding of engineering and information technology fundamentals will grow. All of these gains will increase your confidence to excel in your work and achieve whatever goals you set for yourself. Through this process, you will build your self-efficacy.

You will have ample opportunities to build your self-respect as well. Success on the job will bring positive feedback from your managers and your colleagues. More tangible rewards such as challenging work assignments, leadership roles, and merit salary increases will be yours.

More importantly you will benefit from the satisfaction of doing a good job on projects that will make a difference in the world. You may have opportunities to write papers on your work and present them at local, regional, or national conferences. These accomplishments will be respected by others and will enhance your sense of self-worth.

REFLECTION

Review my top ten list of rewards and opportunities that will come to you if you are successful in getting your engineering degree. Which item on the list is most important to you? Money? Prestige? Challenging work? Making a difference in the world? Reflect on the one you chose. Why did you choose it? Why is that one so important to you?

2.5 ENGINEERING PAST - *GREATEST ENGINEERING ACHIEVEMENTS OF THE 20ᵀᴴ CENTURY*

Although engineering achievements have contributed to the quality of human life for more than 5,000 years [19], the 20ᵗʰ century stands out for its remarkable engineering progress and innovation. In recognition of this, as we entered the 21ˢᵗ century, the National Academy of Engineering launched a project to select the 20 "Greatest Engineering Achievements of the 20ᵗʰ Century."

The primary selection criterion was the impact of engineering achievements on the quality of life in the 20ᵗʰ century. William A. Wulf, president of the National Academy of Engineering, summed it up well:

> *__Engineering__ is all around us, so people often take it for granted, like air and water. Ask yourself, what do I touch that is not engineered? Engineering develops and delivers consumer goods, builds the networks of highways, air and rail travel, and the Internet; mass produces antibiotics; creates artificial heart valves; builds lasers; and offers such wonders as imaging technology and conveniences like microwave ovens and compact discs. In short, engineers make our quality of life possible.*

Following are the "Greatest Engineering Achievements," presented by Neil Armstrong at the National Press Club in Washington, D.C., on February 22, 2000:

    #20 - High Performance Materials
    #19 - Nuclear Technologies
    #18 - Laser and Fiber Optics
    #17 - Petroleum and Gas Technologies
    #16 - Health Technologies
    #15 - Household Appliances
    #14 - Imaging Technologies
    #13 - Internet
    #12 - Space Exploration
    #11 - Interstate Highways
    #10 - Air Conditioning and Refrigeration
    #9 - Telephones
    #8 - Computers
    #7 - Agricultural Mechanization
    #6 - Radio and Television
    #5 - Electronics
    #4 - Safe and Abundant Water
    #3 - Airplanes
    #2 - Automobiles
    #1 - Electrification

Brief descriptions of each are presented in Appendix D. Detailed descriptions of each great achievement can be found on the web at www.greatachievements.org and also in an excellent book titled *A Century of Innovation: Twenty Engineering Achievements that Transformed Our Lives* [1].

REFLECTION

Review the "Greatest Engineering Achievements of the 20th Century." How important are each of these achievements to the quality of our lives? Think about the role of engineers in each of these achievements. Do you think engineers get the credit they deserve for making our lives better? If you don't think they do, why do you think that is?

2.6 ENGINEERING DISCIPLINES

At this point you should have a general understanding of what engineering is and what engineers do – along, of course, with the many rewards and opportunities that engineering offers. Our goal in the remainder of this chapter is to clarify and broaden that understanding. We'll start by looking at engineering from a new perspective: how engineers can be classified by their academic discipline.

Until recently, engineering has consisted of five major disciplines, which graduate the largest number of students. In rank order, these disciplines are:

- **Mechanical engineering**
- **Electrical engineering**
- **Civil engineering**
- **Chemical engineering**
- **Industrial engineering**

A sixth discipline, **computer engineering**, has now been added to this list as has the rapidly growing disciplines of **bioengineering and biomedical engineering,** which as a combined field has passed up industrial engineering as the fifth largest discipline.

Initially a subspecialty within electrical engineering (and still organized that way at many institutions), computer engineering has grown so quickly that institutions are increasingly offering separate accredited B.S. degrees in this field. (Given these changes, computer engineering is treated separately in the discussion of engineering disciplines in Appendix E.)

In addition to the largest traditional disciplines, there are many other more specialized, non-traditional fields of engineering. Aerospace engineering, materials engineering, ocean engineering, petroleum engineering, mining engineering, nuclear engineering, and manufacturing engineering are examples of these.

The following table shows the number of accredited programs in each engineering discipline and degrees awarded in 2010/11 in that discipline. You might note that two-thirds of the B.S. degrees awarded were in electrical (including computer), mechanical, and civil engineering.

To find out which of these engineering programs are offered at each of the 389 institutions in the U.S. that have at least one accredited engineering program, visit the Accreditation Board for Engineering and Technology website at: main.abet.org/aps/accreditedprogramsearch.aspx. There you can search for listings of accredited engineering programs by discipline (e.g., electrical, mechanical, civil, etc.) and by geographical location (region or state).

An overview of each engineering discipline is presented in Appendix E. For the top eight disciplines, more information is provided, while the smaller disciplines are given briefer descriptions.

The page number where you will find the description of each discipline is shown in the table on the next page. If you have already decided on your discipline, I would encourage you to carefully study the information provided about that discipline. If you have not yet decided on your discipline, reviewing the descriptions of all of the disciplines you might be interested in can help you with that decision.

ENGINEERING DISCIPLINES RANKED
BY NUMBER OF B.S. DEGREES AWARDED: 2010/11 [20]

Discipline	Number of Accredited Programs	B.S. Degrees Awarded in 2010/11	Location of Description of Discipline
Mechanical engineering	289	19,016	Page 287
Civil engineering	224	13,175	Page 290
Electrical and electronics engineering	297	12,005	Page 285
Computer engineering	218	11,610	Page 292
Chemical engineering	158	6,297	Page 296
Bioengineering and biomedical engineering	73	4,293	Page 297
Industrial engineering	93	3,423	Page 298
Aerospace engineering	65	3,286	Page 299
General engineering/engineering physics/engineering science	73	2,812	–
Materials engineering/metallurgical engineering	65	1,134	Page 299
Petroleum engineering	17	994	Page 301
Environmental engineering	59	740	Page 301
Architectural engineering	17	730	Page 300
Naval architecture/marine engineering/ocean engineering	16	578	Page 301
Engineering management	11	504	Page 303
Systems engineering	16	472	Page 300
Nuclear and radiological engineering	20	463	Page 302
Agricultural engineering/biological engineering	47	427	Page 300
Mining engineering/metallurgical engineering/geological engineering	35	417	Page 302
Manufacturing engineering	21	146	Page 302
Ceramic engineering	3	52	Page 302
Software engineering	21	**	Page 302
Construction engineering	12	**	Page 303
Optics engineering	5	**	Page 303
Surveying and geomatics engineering	5	**	Page 303

Engineering mechanics	5	**	–
Telecommunications engineering	2	**	Page 304
Welding engineering	1	**	–
Fire protection engineering	1	**	–
Other	–	2,014	–
TOTAL	**1,885**	**84,599**	–

** - Included in disciplines with degrees awarded listed as "Other"

> ### REFLECTION
> Reflect on the 25 engineering disciplines described in Appendix E. Have you already decided which one you will major in? Why did you choose it? If you haven't yet chosen a specific engineering discipline, which one is the most appealing to you at this point? What about it do you find appealing?

2.7 ENGINEERING JOB FUNCTIONS

As noted earlier in this chapter, another way to understand the engineering profession is to examine engineers from the perspective of the work they do or the job functions they perform. For example, an electrical engineer could also be referred to as a *design* engineer, a *test* engineer, or a *development* engineer – depending on the nature of his or her work. Following is a description of the ten main engineering job functions.

ANALYSIS

The **analytical engineer** is primarily involved in the mathematical modeling of physical problems. Using the principles of mathematics, physics, and engineering science – and making extensive use of engineering applications software – the analytical engineer plays a critical role in the initial stage of a design project, providing information and answers to questions that are easy and inexpensive to obtain. Once the project moves from the conceptual, theoretical stage to the actual fabrication and implementation stage, changes tend to be time-consuming and costly.

DESIGN

The **design engineer** converts concepts and information into detailed plans and specifications that dictate the development and manufacture of a product. Recognizing that many designs are possible, the design engineer must consider such factors as production cost, availability of materials, ease of production, and performance requirements. Creativity and innovation, along with an analytic mind and attention to detail, are key qualifications for a design engineer.

TEST

The **test engineer** is responsible for developing and conducting tests to verify that a selected design or product meets all specifications. Depending on the product, tests may be required for such factors as structural integrity, performance, or reliability – all of which must be performed under all expected environmental conditions. Test engineers also conduct quality control checks on existing products.

DEVELOPMENT

The **development engineer**, as the title indicates, is involved in the development of products, systems, or processes. The context in which such development occurs, however, can vary considerably. Working on a specific design project, the development engineer acts as a kind of intermediary between the design and test engineers. He helps the design engineer formulate as many designs as possible that meet all specifications. Once a design is selected, the development engineer oversees its fabrication – usually in the form of a prototype or model. The results of his collaboration with the design engineer and subsequent supervision of the prototype's fabrication are bound to affect the kind and amount of testing the test engineer will then need to conduct.

In a more general context, the development engineer is instrumental in turning concepts into actual products or applying new knowledge to improve existing products. In this capacity, she is the "D" in "R&D," which, as you probably know, stands for research and development. Here, the development engineer is responsible for determining how to actualize or apply what the researcher discovers in the laboratory, typically by designing, fabricating, and testing prototypes or experimental models.

SALES

The **sales engineer** is the liaison person between the company and the customer. In this role, the sales engineer must be technically proficient in order to understand the product itself and the customer's needs. That means he must be able to explain the product in detail: how it operates, what functions it can perform, and why it will satisfy the customer's requirements. He also needs to maintain a professional working relation-

ship as long as the customer is using his company's products. He must be able to field questions about the product, explain its features to new users, and arrange prompt service should the customer experience problems with the product. Obviously, along with solid technical knowledge, the sales engineer must possess strong communication skills and "people" skills.

RESEARCH

The work of the **research engineer** is not unlike that of a research scientist in that both are involved in the search for new knowledge. However, the difference lies in what motivates their work. Scientific researchers are generally interested in the new knowledge itself: what it teaches or uncovers about natural phenomena. Engineering researchers are interested in ways to <u>apply</u> the knowledge to engineering practices and principles. Research engineers thus explore mathematics, physics, chemistry and engineering sciences in search of answers or insights that will contribute to the advancement of engineering.

Given the nature and demands of their work, research engineers usually need to have an advanced degree. Indeed, most positions available in engineering research require a Ph.D.

MANAGEMENT

If you are successful as an engineer and have strong leadership skills, within a few years of graduation you could very well move into management. Opportunities exist primarily in two areas: **line management** and **project management**.

In a company, the technical staff is generally grouped into a "line organization." At the base of this "line" are units of ten to 15 engineers, managed by a unit supervisor. At the next level up the line, these units report to a group manager. This organizational line continues up to department managers, a chief engineer or engineering vice president, and finally the president. Often the president of a technical company is an engineer who worked his or her way up through the line organization.

Project management is a little different, as the personnel are organized according to a specific project or assignment. At the head of each project is a project manager. For a small project, one manager is usually sufficient to oversee the entire project; for a larger project, the project manager is assisted by a professional staff, which can range from one to several hundred people. The overall responsibility of the project manager and staff is to see that the project is completed successfully, on time, and within budget.

CONSULTING

The work of the **consulting engineer** differs from that of all other engineers in that a consulting engineer performs services for a client on a contractual basis. Some consulting engineers are self-employed, while others work for consulting firms that hire out their engineers to companies that either lack the expertise the consulting engineer can provide or want an outside evaluation of their organization's performance. Depending on the client's specific needs, the consulting engineer's work can vary considerably. Investigations and analyses; preplanning, design and design implementation; research and development; construction management; and recommendations regarding engineering-related problems are just a few examples.

The time a consulting engineer puts into each assignment also can vary. Sometimes the work can be done in a day; other times it can require weeks, months, or even years to complete.

Last, engineering consulting is increasingly becoming a global enterprise. Both the public and private sectors of developing nations have growing technological needs and turn to U.S. consulting firms for technical assistance. If the diversity of work and the opportunity to travel catch your interest, a career in engineering consulting could be for you.

TEACHING

The **engineering professor** has three primary areas of responsibility: teaching, research, and service. Teaching includes not only classroom instruction, but also course and curriculum development, laboratory development, and the supervision of student projects or theses. Research involves the pursuit of new knowledge, which is then disseminated throughout the professional engineering community by papers published in engineering journals, presentations at scholarly meetings, textbooks, and software.

The research demands of the engineering educator also include success in generating funds to support research projects as well as participation in professional societies.

"Service" is a catch-all term that refers to the many functions expected of engineering professors. These include such activities as community involvement, participation in faculty governance, public service, and consulting.

The Ph.D. degree in engineering is virtually mandatory to qualify for a full-time position on an engineering faculty at a four-year institution, while an M.S. degree is generally sufficient for a teaching position at a community college. More information about academic careers in engineering can be found in Reference 21.

ENTREPRENEUR

An *entrepreneur* is a person who starts a new business venture. If you are able to combine your strong technical expertise with qualities of inventiveness, risk-taking, and a sense of adventure, someday you may be the president of your own company. There are many examples of engineer entrepreneurs. Some of the most prominent examples are William Hewlett and David Packard (Hewlett-Packard), Jeff Bezos (Amazon.com), Pierre Omidyar and Jeffrey Skoll (E-Bay), Larry Page and Sergey Brin (Google), Andrew Grove (Intel), Leonard Bosack (Cisco Systems), and Jack Gifford (AMD). And these are not exceptions. A 2009 report by the Kaufman Foundation [22] estimated that slightly more than 17,000 companies have been founded by living MIT engineering graduates.

The fact that many engineers become entrepreneurs should come as no surprise. Engineers develop products or processes to meet desired needs. The creative and analytical skills of an engineer can also be put to use in developing a business, particularly one that focuses on high-tech products or services. A list of the traits of successful entrepreneurs would include disciplined, confident, open-minded, a self-starter, competitive, creative, determined, strong work ethic, strong people skills, passionate. If you combine these traits with a strong technical background, you may someday be the CEO of your own company.

The opportunity for engineers to start businesses is on the increase as technology plays a more and more important role in our lives. One reflection of this is that a growing number of universities are developing courses or even entire programs on entrepreneurship for engineers.

REFLECTION

Consider the ten engineering job functions described in this section. Which of them appeals to you? Analysis? Design? Test? Development? Sales? Research? Could you see yourself in management? Could you see yourself as a consulting engineer? How about being an engineering professor or entrepreneur?

2.8 EMPLOYMENT OPPORTUNITIES

When you graduate in engineering, you will face a number of choices. The first will be whether you want to continue your education full time or seek work as a practicing engineer. If you elect to continue your education, you next need to decide whether you want to pursue an M.S. degree in engineering or do graduate work in another field, such as business administration, law, or medicine. (Opportunities for graduate study are discussed in Chapter 8.)

If you decide to seek a full-time engineering position, many opportunities and choices await you. The field of engineering practice is so vast and the job opportunities so varied, you may well need to devote a substantial amount of time to fully understand the opportunities and areas of practice available to you.

Rather than waiting until you graduate to learn about these many opportunities, you should make this an objective early on in your engineering studies. Besides saving time and energy when you launch your job search later, knowing NOW about the many areas in which engineers are needed and the diverse opportunities that await you will be a strong incentive for you to complete your engineering studies.

Let's start, then, with a "big picture" view of the major areas in which most engineers work. The table below, which lists these areas, along with the percentages in each [23]:

Employed Individuals with Engineering Degrees

Employment Area	Percentage
Business/Industry	80.3%
Federal Government	5.4%
State/Local Government	5.7%

Educational Institutions	5.1%
Self-Employed	3.5%
Total	**100%**

As you can see, the first area, "Business/Industry," is clearly the largest, employing 80.2 percent of engineers. You should know, however, that "industry" is a blanket term for two distinct categories: (1) manufacturing, and (2) non-manufacturing (or service). *Manufacturing* is involved in converting raw materials into products, while *non-manufacturing* concerns the delivery of services. Government, the next highest area, employing 11.1 percent of engineers, has needs for engineers at the local, state, and federal levels. Following business, industry, and government come educational institutions, which employ 5.1 percent of engineers, both as engineering professors and as researchers in university-operated research laboratories. Finally, there is a small but significant area of self-employed engineers, most of whom are consulting engineers.

ORGANIZATION OF INDUSTRY IN THE UNITED STATES

If 80 percent of engineers work in business and industry, it is likely that you, too, will find yourself working in this area. Although we briefly mentioned the two categories into which industry is divided (manufacturing and non-manufacturing), we have barely scratched the surface of this huge, complex field. For a comprehensive look at the many diverse fields that comprise

U.S. business and industry, the **2012 North American Industry Classification System (2012 NAICS)** [24] is the best resource available.

Developed and maintained by the U.S. government, the 2012 NAICS system dissects the monolithic term "business and industry" into 1,067 "national industries," each identified by a six-digit classification code. It then lists all the products or services that each national industry provides.

To give you an idea of how the 2012 NAICS works, I randomly selected ten of the 1,067 national industries in the NAICS classification system:

211111   Crude petroleum and natural gas extraction
221112   Electric power generation, fossil fuel (e.g., coal, oil, gas)
237310   Highway, street, and bridge construction
325611   Soap and other detergent manufacturing
334510   Electromedical and electrotherapeutic apparatus manufacturing
335311   Power, distribution, and specialty transformer manufacturing
335921   Fiber-optic cable manufacturing
336414   Guided missile and space vehicle manufacturing
517210   Wireless telecommunications carriers (except satellite)
541330   Engineering services

The first two digits designate a major "Economic Sector," and the third digit designates an "Economic Subsector." For example, all of the industry subgroups above starting with 33 are part of the "Manufacturing" Economic Sector. The two national industries in the list whose first three digits are 335 fall under the "Electrical Equipment, Appliance, and Component

Manufacturing" Economic Subsector. The remaining digits of each six-digit classification code further subdivide the subsectors into industry groups (4 digits), NAICS industries (5 digits), and national industries (6 digits).

As an example, consider the Economic Sector, "33 – Manufacturing." Under this economic sector, there are eight "Economic Subsectors":

331 – Primary metal manufacturing
332 – Fabricated metal product manufacturing
333 – Machinery manufacturing
334 – Computer and electronic product manufacturing
335 – Electrical equipment, appliance, and component manufacturing
336 – Transportation equipment manufacturing
337 – Furniture and related product manufacturing
339 – Miscellaneous manufacturing

Take one of the "Economic Subsectors," say NAICS 334, "Computer and electronic product manufacturing." Under this Economic Subsector, there are six "Industry Groups":

3341 – Computer and peripheral equipment manufacturing
3342 – Communications equipment manufacturing
3343 – Audio and video equipment manufacturing
3344 – Semiconductor and other electronic component manufacturing
3345 – Navigational, measuring, electromedical, and control instruments manufacturing
3346 – Manufacturing and reproducing magnetic and optical media

Now, take one of the "Industry Groups," NAICS 3345, for example: "Navigational, measuring, electromedical, and control instruments manufacturing." Under this industry group there are nine national industries:

334510 – Electromedical and electrotherapeutic apparatus manufacturing
334511 – Search, detection, navigation, guidance, aeronautical, and nautical system and instrument manufacturing
334512 – Automatic environmental control manufacturing for residential, commercial, and appliance use
334513 – Instruments and related products manufacturing for measuring, displaying, and controlling industrial process variables
334514 – Totalizing fluid meter and counting device manufacturing
334515 – Instrument manufacturing for measuring and testing electrical signals
334516 – Analytical laboratory instrument manufacturing
334517 – Irradiation apparatus manufacturing
334519 – Other measuring and controlling device manufacturing

If you pick just one of these nine national industries, for instance 334510 – Electromedical and electrotherapeutic apparatus manufacturing – you'll find listed more than 50 major product groups such as magnetic resonance imaging equipment, medical ultrasound equipment, pacemakers, hearing aids, electrocardiographs, and electromedical endoscopic equipment.

I hope this has given you an idea of the enormity of U.S. business and industry and the tools you need to access that industry. You can explore the North American Industry Classification System on your own online at www.census.gov/naics. Under "2012 NAICS Search," enter one of two types of keywords:

(1) a product or service (e.g., "fiber optic")
(2) a two- to six-digit NAICS classification (e.g., "541330")

We learned earlier that 80 percent of engineers work in business and industry. The following sections briefly describe the economic subsectors that employ the largest number of engineers in both the manufacturing subsector and the service sector.

MANUFACTURING SUBSECTORS

Below are brief descriptions of the six manufacturing subsectors employing the largest numbers of engineers.

COMPUTER AND ELECTRONIC PRODUCT MANUFACTURING. These industries are engaged in the manufacture of computers, computer peripherals, communication equipment, and related electronic equipment. Their manufacturing processes differ fundamentally from those of other machinery and equipment in that the design and use of integrated circuits and the application of highly specialized miniaturization technologies are common elements in the manufacturing processes of computer and electronic products.

TRANSPORTATION EQUIPMENT MANUFACTURING. These industries produce equipment and machinery needed for transporting people and goods. Their manufacturing processes are similar to those used in most other machinery manufacturing establishments: bending, forming, welding, machining, and assembling metal or plastic parts into components and finished products. Evidence of the equipment and machinery manufactured in this subsector can be found in a variety of products - motor vehicles, aircraft, guided missiles and space vehicles, ships, boats, railroad equipment, motorcycles, bicycles, and snowmobiles.

MACHINERY MANUFACTURING. These industries design and produce products that require mechanical force to perform work. Both general-purpose machinery and machinery designed to be used in a particular industry are included in this subsector. Examples of general-purpose machinery include heating, ventilation, air-conditioning, and commercial refrigeration equipment; metalworking machinery; and engine, turbine, and power transmission equipment. Special-purpose machinery are included: agricultural, construction, and mining machinery; industrial machinery; and commercial and service industry machinery.

FABRICATED METAL PRODUCT MANUFACTURING. These industries transform metal into intermediate or end products using forging, stamping, bending, forming, and machining to shape individual pieces of metal. They also use processes, such as welding and assembling, to join separate parts together. Examples of products include hand tools, kitchen utensils, metal containers, springs, wire, plumbing fixtures, firearms, and ammunition.

CHEMICAL MANUFACTURING. These industries manufacture three general classes of products: (1) basic chemicals, such as acids, alkalies, salts, and organic chemicals; (2) chemical products to be used in further manufacture, such as synthetic fibers, plastics materials, dry colors, and pigments; and (3) finished chemical products to be used for human consumption, such as drugs, cosmetics, and soaps; or products to be used as materials or supplies in other industries, such as paints, fertilizers, and explosives.

ELECTRICAL EQUIPMENT, APPLIANCE, AND COMPONENT MANUFACTURING.
These industries manufacture products that generate, distribute, and use electrical power. Electric lighting equipment, household appliances, electric motors and generators, batteries, and insulated wire and wiring devices are but a few of the many products that come under this manufacturing subsector.

SERVICE SECTORS

The following provides brief descriptions of the eight service sectors that employ the greatest number of engineers.

PROFESSIONAL, SCIENTIFIC, AND TECHNICAL SERVICES. This sector includes industries from three large areas, only one of which – "Technical Services" – applies to engineering. Under "Technical Services," however, NAICS includes a broad, varied list of both engineering and computer services. Engineering services may involve any of the following: provision of advice

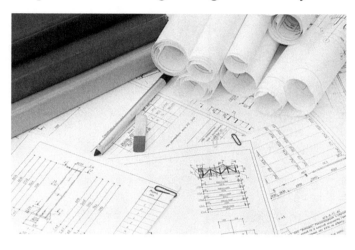

(i.e., engineering consulting), preparation of feasibility studies, preparation of preliminary plans and designs, provision of technical services during the construction or implementation stages of a project, inspection and evaluation of completed projects, and related services. Computer services are equally varied, including activities such as programming, computer-integrated systems design, data preparation and processing, information retrieval, facilities management, as well as computer leasing, maintenance, and repair.

INFORMATION. These industries focus on three main processes: (1) producing and distributing information and cultural products; (2) providing the means to transmit or distribute these products, along with data or communications; and (3) processing data. Subsectors include publishing industries, motion picture and sound recording industries, broadcasting and telecommunications, and information and data processing services.

CONSTRUCTION. These industries cover three broad areas of construction: (1) building construction, such as dwellings, office buildings, commercial buildings, stores, and farm buildings; (2) heavy construction other than buildings, such as highways, streets, bridges, sewers, railroads, irrigation projects, flood control projects, and marine construction; and (3) special trades for heavy construction such as painting, electrical work, plumbing, heating, air-conditioning, roofing, and sheet metal work.

WHOLESALE TRADE. Wholesale trade includes: (1) merchant wholesalers who take title to the goods they sell; (2) sales branches or offices maintained by manufacturing, refining, or mining enterprises; and (3) agents, merchandise or commodity brokers, and commission merchants. The merchandise includes the output of agriculture, mining, manufacturing, and certain information industries, such as publishing.

ADMINISTRATION AND SUPPORT. These service industries perform routine support activities for the day-to-day operations of other organizations. These essential activities are often undertaken in-house by establishments in many sectors of the economy. Activities include office administration, personnel employment and placement, document preparation and other clerical services, solicitation, collection, security and surveillance, and waste disposal.

MANAGEMENT OF COMPANIES AND ENTERPRISES. This sector involves (1) holding securities of (or other equity interests in) companies and enterprises for the purpose of owning a controlling interest or influencing management decisions, or (2) administering, overseeing, and managing companies. Engineers who are provided to companies on a contract basis are included in this economic sector.

UTILITIES. These industries are engaged in providing electric power, natural gas, steam, water, and sewage removal. Providing electric power includes generation, transmission, and distribution, while natural gas only involves distribution. Supplying steam includes provision and/or distribution; supplying water involves treatment and distribution. Sewage removal includes collection, treatment, and disposal of waste through sewer systems and sewage treatment facilities.

MINING. These industries extract naturally occurring mineral solids, such as coal and ores; liquid minerals, such as crude petroleum; and gases, such as natural gas. The term "mining" is used in the broad sense to include quarrying, well operations, beneficiating (e.g., crushing, screening, washing, and flotation), and other preparatory functions customarily done at the mine site.

2.9 IMPORTANT FIELDS FOR THE FUTURE

We are in a period of intense change. One way to underscore this is to reflect on the fact that none of the 50 technological inventions listed below existed as recently as 1982 (30 years ago).

50 GREATEST TECHNOLOGICAL INVENTIONS OF PAST 25 YEARS

50	Hybrid cars	32	Bluetooth
49	Mini disc	31	Commercialized GPS
48	Color plasma display	30	Home audio editing
47	Optical computer mouse	29	Home video editing
46	LED headlights	28	Camcorders
45	Electronic tolls	27	Digital SLR cameras
44	OLED TV	26	Multi-core processors
43	Blu-ray	25	Satellite radio
42	Satellite TV	24	Flip phones
41	Recordable DVDs	23	Digital HDTV
40	Lithium rechargeable batteries	22	Instant messaging
39	DVD	21	Consumer digital cameras
38	CD-R	20	JPEG
37	Voice mail	19	Microblogging
36	Online stock trading	18	Caller ID
35	Doppler radar	17	Mobile broadband
34	MPEG-4	16	Blogs
33	Flash memory	15	MP3 players

14	Electronic word processing	13	DVR
12	DNA profiling	6	Text messaging
11	Social networking service	5	Wi-Fi
10	Genetic sequencing	4	MP3
9	Web-based email	3	Broadband Internet
8	Search engines	2	Personal computers
7	Smart phones	1	World Wide Web

This list was compiled by *Complex Magazine* and published in August, 2010 [26]. Not only have these inventions changed our ways of communicating and locating information, they have created exciting new opportunities for engineers. The ranking is based on which inventions affect our lives the most.

REFLECTION

Review the "50 Greatest Technological Inventions of the Past 25 Years." Which items will help you reach your goal of becoming an engineer? How so? More importantly, which ones have the potential to interfere with your goal of becoming an engineer? What can you do to ensure that they don't keep you from achieving your goal of becoming an engineer?

Many of the inventions listed above have resulted in a "flattening" of the world. To see what this means, check out Thomas L. Friedman's profound book *The World Is Flat: A Brief History of the Twenty-First Century* [25]. Doing so will help you understand the world in which you are living and working. Friedman explores the political and technological changes that have "flattened" the world, making it a smaller place. Such events such as the fall of the Berlin Wall, the explosion of the Internet, the dot-com bubble and bust, the outsourcing of jobs to India and China, and "globalization" have leveled the playing field for many emerging economies.

The following is a list of some of the major events and changes that will influence your future as an engineer.

Major Events and Changes Affecting the Future

Fall of the Berlin Wall
Advances in computer technology
Advances in communications
The knowledge and information explosion
Globalization (outsourcing, off-shoring)
Environmental challenges/sustainability
World population explosion
Democratization
Pandemic diseases/drug resistant germs
Climate change/natural disasters
Nuclear proliferation
Events of September 11, 2001/threat of terrorism

Future Directions - *Grand Challenges for Engineering*

A good way to understand future directions for engineering is to examine the National Academy of Engineering's "Grand Challenges for Engineering." The 14 challenges, announced in 2008, were developed by a select committee with input from broad constituencies throughout the world. They address four themes considered "essential for humanity to flourish:" environmental sustainability, health, reducing our vulnerability, and adding to the joy of living.

GRAND CHALLENGES FOR ENGINEERING

1. Make solar energy economical.
2. Provide energy from fusion.
3. Develop carbon sequestration methods.
4. Manage the nitrogen cycle.
5. Provide access to clean water.
6. Restore and improve urban infrastructure.
7. Advance health informatics.
8. Engineer better medicines.
9. Reverse-engineer the brain.
10. Prevent nuclear terror.
11. Secure cyberspace.
12. Enhance virtual reality.
13. Advance personalized learning.
14. Engineer the tools of scientific discovery.

Understanding these challenges can help you prepare for the engineering fields that will be particularly important in the years ahead. The following sections briefly discuss each challenge. Detailed information on each can be found at www.engineeringchallenges.org.

MAKE SOLAR ENERGY ECONOMICAL. Sunshine has long offered a tantalizing source of environmentally-friendly power, bathing the earth with more energy each hour than the planet's population consumes in a year. But capturing that power, converting it into useful forms, and storing it for the proverbial rainy day, pose provocative engineering challenges.

PROVIDE ENERGY FROM FUSION. Nuclear fusion, the artificial re-creation of the sun's source of power on Earth, offers the potential for long-term energy supply. Human-engineered fusion has already been demonstrated on a small scale. The challenges facing the engineering community are to find ways to scale up the fusion process to commercial proportions, in an efficient, economical, and environmentally friendly way.

DEVELOP CARBON SEQUESTRATION METHODS. Engineering solutions for solar power and nuclear fusion must be feasible not only technologically but also economically when compared with fossil fuels. Even with success, however, it remains unlikely that fossil fuels will be eliminated from the planet's energy budget anytime soon, leaving their impact on the environment for engineers to address. Most notoriously, evidence is mounting that the carbon dioxide pumped into the air by burning fossil fuels is increasing the planet's temperature and threatens to have disruptive effects on climate. Consequently, engineers will need to find ways of capturing and sequestering the carbon dioxide produced from fuel burning.

MANAGE THE NITROGEN CYCLE. A lesser known environmental concern involves nitrogen, the atmosphere's most abundant component. The natural cycle that extracts nitrogen from the air for its incorporation into plants – and hence food – has become altered by human activity. With widespread use of fertilizers and high-temperature industrial combustion, humans have doubled the rate at which nitrogen is removed from the air since pre-industrial times, contributing to smog and acid rain, polluting drinking water, and even worsening global warming. Engineers

must design countermeasures for nitrogen-cycle problems, while maintaining the ability of agriculture to produce adequate food supplies.

PROVIDE ACCESS TO CLEAN WATER. The short supply of quality water for personal use – drinking, cleaning, cooking, and removal of waste – and large-scale use such as irrigation for agriculture, is a vital concern in many regions of the world. Engineers will play a role in meeting water needs through both the desalination of sea water and also the use of small-scale technologies for local water purification.

RESTORE AND IMPROVE URBAN INFRASTRUCTURE. America's infrastructure – water and sewer systems, hazardous and solid waste disposal systems, roads and bridges, rail networks, urban transportation systems, power and natural gas grids, and public buildings – is aging and failing. Engineers face the enormous challenge of renewing and sustaining these infrastructures while preserving ecological balances and enhancing the aesthetic appeal of living spaces.

ADVANCE HEALTH INFORMATICS. As computers have become critical to so many facets of human endeavors, there is now a consensus that a computer-based approach to health informatics – the acquisition, management, and use of information in health – can greatly enhance the quality and efficiency of medical care delivery and the response to widespread public health emergencies. Health and biomedical informatics encompass issues from the personal to the global, ranging from medical records for individual patients to sharing data about disease outbreaks. Maintaining a healthy world population in the 21st century will require systems engineering approaches to redesign care practices and integrate local, regional, national, and global health informatics networks.

ENGINEER BETTER MEDICINES. Doctors have long recognized that individuals differ in their susceptibility to disease and their response to treatments, but medical technologies have generally been offered as "one size fits all." Recent cataloging of the human genome, along with a better understanding of the body's complement of proteins and their biochemical interactions, offer the prospect of identifying the specific factors that determine sickness and wellness in any individual. Examples of engineering challenges are developing better systems to quickly assess a patient's genetic profile, collecting and managing massive amounts of data on individual patients, and creating new diagnostic devices such as gene chips and sensors able to detect minute amounts of chemicals in the blood.

REVERSE ENGINEER THE BRAIN. While some machines have mastered specific narrow thinking skills – playing chess, for instance – general-purpose artificial intelligence (AI) has remained elusive. Part of the problem, some experts now believe, is that artificial brains have been designed without enough attention to real ones. The secrets about how living brains work are bound to offer the best guide to engineering an artificial variety. Discovering those secrets by reverse-engineering the brain promises enormous opportunities for reproducing intelligence the way assembly lines turn out cars or computers.

PREVENT NUCLEAR TERROR. Since the beginning of the Nuclear Age, the materials needed to make a nuclear weapon have been accumulating around the world. Worse yet, the instructions for building explosive devices from such materials have been widely published, suggesting that access to the ingredients would make a bomb a realistic possibility.

Challenges for engineers include: (1) how to secure the materials, (2) how to detect hidden nuclear materials, (3) how to render a potential device harmless, (4) how to deal with the aftermath of a nuclear explosion, and (5) how to determine who is responsible for an attack.

Vulnerability to biological disaster is also of great concern, and technologies for early detection of such threats and rapid deployment of countermeasures (such as vaccines and antiviral drugs) rank among the most urgent of today's engineering challenges.

SECURE CYBERSPACE. It is clear that engineers need to develop solutions for a long list of cybersecurity problems. For one, better approaches are needed to authenticate hardware, software, and data in computer systems and to verify user identities. Biometric technologies, such as fingerprint readers, may be one step in that direction.

Engineering more secure software is a particularly critical need. One way to do this may be through better programming languages that have security protection built into the ways programs are written. Another challenge is to ensure that data transmitted over various routes on the Internet cannot be diverted, monitored, or altered.

ENHANCE VIRTUAL REALITY. For virtual reality systems to effectively simulate reality, several engineering hurdles must be overcome. The resolution of the video display must be high, with rapid refresh and update rates, for scenes to look like and change as they do in real life. The field of view must be wide enough and the lighting and shadows realistic enough to maintain the illusion of a real scene. And for serious simulations, reproducing sensations of sound, touch, and motion is necessary.

ADVANCE PERSONALIZED LEARNING. The external world is not the only place where engineering matters; the inner world of the mind can also benefit from improved methods of instruction and learning. Ways must be found to tailor the mind's growth to an individual's propensities and abilities. New methods of instruction, such as computer-created virtual realities, also apply to entertainment and leisure, furthering engineering's contributions to the joy of living.

ENGINEER THE TOOLS OF SCIENTIFIC DISCOVERY. The spirit of curiosity, whether in individual minds and in society as a whole, can be further promoted through engineering endeavors enhancing exploration at the frontiers of reality and knowledge. New tools are needed for investigating the vastness of the cosmos or the inner intricacy of life and subatomic particles.

2.10 SUSTAINABILITY

Sustainability is the capacity to endure. While many people disagree on exactly what that means, perhaps the best definition was offered by a commission of the United Nations in 1987:

> ***Sustainability** is meeting the needs of the present without compromising the ability of future generations to meet their own needs.*

I'm sure no one would disagree that this definition puts forth a worthy goal, but it is one we are a long way from achieving. Examples of major environmental problems that will be passed to future generations are:

- Global warming/climate change
- Ozone depletion
- Water quality and quantity
- Air pollution
- Dependence on fossil fuels/energy crisis
- Unsustainable agriculture
- Threat of disease
- Waste management and land pollution
- Over-consumption
- World hunger
- Loss of ecosystems/deforestation/animal extinction

These problems are the unintended consequences of rapid population growth, economic growth, and unbridled consumption of natural resources.

Although solutions to these problems involve complex interactions among economic, societal, and environmental factors, engineers will play key roles in achieving sustainability. It is no surprise that a number of the Grand Challenges for Engineering include making solar energy economical, providing energy from fusion, developing carbon sequestration methods, managing the nitrogen cycle, and providing access to clean water directly – all of which address one or more of these problems.

The role of the engineering design process in sustainability is underscored by the fact that one of the attributes of new engineering graduates required by ABET is:

> *an ability to design a system, component, or process to meet desired needs within realistic constraints such as economic, environmental, social, political, ethical, health and safety, manufacturability, and **sustainability***

The work of engineers in solving the environmental problems listed above is often called "green engineering." Green engineering focuses on the design of materials, processes, systems, and devices with the objective of minimizing overall environmental impact (including energy utilization and waste production) throughout the entire life cycle of a product or process.

Some of the important attributes of green engineering and sustainable designs are:

- Designs that use less energy or reduce emission

- Designs with minimal carbon footprints
- Designs that reduce material usage or waste in manufacturing
- Designs with no toxic materials
- Designs that comply with environmental standards and regulations
- Manufacturing processes that use less energy and natural resources
- Products that can be disposed of safely, including biodegradable materials and packaging
- Manufacturing processes that minimize the usage or production of substances of concern
- Designs that use renewable/recyclable/recycled materials
- Products that require less packaging

As a future engineer, you can help in answering the question: "How can engineering practice and technology help to move products, processes, and systems toward sustainability?" To this end, you will have limitless opportunities to be involved in technologies that use less energy, tap fewer natural resources, avoid polluting the environment, and are reusable. Needless to say, the need for creative and innovative thinking will be greater than ever – a reality best captured by the following prescient quote from Albert Einstein:

> *The significant problems we face cannot be solved at the same level of thinking we were at when we created them.*

2.11 ENGINEERING AS A PROFESSION

When you receive your B.S. degree in engineering, you will join the engineering profession. Engineering is considered a profession in that it meets the following characteristics of a learned professional group [27]:

- Knowledge and skill in specialized fields above that of the general public
- A desire for public service and a willingness to share discoveries for the benefit of others
- Exercise of discretion and judgment
- Establishment of a relation of confidence between the professional and client or employer
- Self-imposed (i.e., by the profession) standards for qualifications
- Acceptance of overall and specific codes of conduct
- Formation of professional groups and participation in advancing professional ideals and knowledge
- Recognition by law as an identifiable body of knowledge

REFLECTION

Review the above eight characteristics of a learned profession. Do they describe something you would like to be part of? Which characteristic would you welcome? Are there any that you would have a problem with? Do you have a desire for public service? Are you willing to share what you know to benefit others? Would you look forward to establishing a relation of confidence between you and your employer? Would you welcome the opportunity to learn about and practice codes of conduct? Would you enjoy being part of a profession that requires a great deal of discretion and judgment?

As an engineering professional, you will have certain rights and privileges, as well as certain responsibilities and obligations. As described above, you will be responsible for serving the public, sharing your discoveries for the benefit of others, exercising discretion and judgment, maintaining confidentiality with clients and employers, and abiding by specific codes of conduct.

As an engineering professional, you will have the legal right to represent yourself using the title of engineer. You will be eligible to participate in professional organizations. And you will have the right to seek registration as a *Professional Engineer.*

PROFESSIONAL REGISTRATION

You can formalize your status as a professional by seeking registration as a *Professional Engineer* (P.E.). Professional registration is an impressive credential, and you will find the title **P.E.** proudly displayed on the business cards of engineers who have acquired that status. For most engineers, professional registration is optional. However, in certain fields of work that involve public safety, professional registration may be mandatory. Approximately 30 percent of all practicing engineers are registered. The percentage is much higher for civil engineers because of the nature of their work.

Professional registration is handled by the individual states, each of which has a registration board. Although the requirements and procedures differ somewhat from state to state, they are generally fairly uniform due to the efforts of the National Council of Examiners for Engineers and Surveyors (NCEES). For details about becoming a registered Professional Engineer, visit the NCEES webpage: www.ncees.org.

State boards are responsible for evaluating the education and experience of applicants for registration, administering an examination to those applicants who meet the minimum requirements, and granting registration to those who pass the exam.

Although registration laws vary, most boards require four steps:

1. Graduation from a four-year engineering program accredited by the Accreditation Board for Engineering and Technology (ABET)
2. Passing the Fundamentals of Engineering (FE) examination
3. Completing four years of acceptable engineering practice
4. Passing the Principles and Practice of Engineering (PE) examination

Once you complete these four steps, you will become licensed as a Professional Engineer in the state in which you wish to practice, and you will be certified to use the prestigious P.E. designation after your name. Most states provide for reciprocal licensure, so that once you become licensed in one state, you can become licensed in other states without further examination.

THE FUNDAMENTALS OF ENGINEERING EXAM. The Fundamentals of Engineering Exam (FE) is administered each year in April and October. It is an eight-hour, closed-book, multiple-choice exam. The only reference material that can be used is the *FE Supplied-Reference Handbook* [28] that will given to you to use during the exam but can be obtained and reviewed beforehand. The four-hour morning session is the same for all engineering disciplines and consists of 120 one-point questions on the following topics:

Mathematics (15%)

Engineering probability and statistics (7%)

Chemistry (9%)

Engineering economics (8%)

Computers (7%)

Ethics and business practice (7%)

Engineering mechanics (statics and dynamics) (10%)

Strength of materials (7%)

Material properties (7%)

Fluid mechanics (7%)

Electricity and magnetism (9%)

Thermodynamics (7%)

The four-hour afternoon exam is comprised of 60 two-point questions and covers one of seven engineering disciplines (electrical, mechanical, civil, industrial, chemical, environmental, general) chosen by you.

The FE exam can be taken prior to graduation, ideally sometime in your senior year or soon after you graduate. This is the time when you have the best command of engineering fundamentals. Once you have passed this exam and graduated, you are designated as an Intern-Engineer or Engineer-in-Training.

One note. Your engineering curriculum may not include courses that cover all of the topics included on the FE exam, so you may need to take an FE review course, do extensive self-study, or even elect to take an additional course or two during your undergraduate years.

Beginning in January 2014, the FE exam will be offered via computer-based testing (CBT) rather than a paper-and-pencil exam. Testing will be administered through Pearson Virtual University Enterprises (VUE) network of testing centers worldwide and will be offered periodically throughout the year rather than on two specific dates. You should expect changes in the exam format. Some of those under consideration include shortening the length of the test, integrating the common and discipline-specific sections, and broadening the types of question formats. You can find information on the status of the change to computer-based testing at: cbt.ncees.org.

THE PRINCIPLES AND PRACTICE OF ENGINEERING EXAM. After four years of experience as an Intern Engineer or Engineer-in-Training, you will be eligible to take the Principles and Practice of Engineering Exam (PE exam). Offered in April and October, the PE exam is an eight-hour, open-book exam in a specific engineering discipline (civil, mechanical, electrical, chemical, industrial, etc.). The exam consists of 40 multiple-choice questions in the four-hour morning session, and 40 multiple-choice questions in the four-hour afternoon session. If you take the exam in the civil, electrical, or mechanical engineering disciplines, the afternoon session will focus on a sub-specialty of the discipline selected by you.

The NCEES plans to transition the PE exam to computer-based testing at some point in the near future, but no sooner than 2015.

PROFESSIONAL SOCIETIES

Each of the engineering disciplines described in Appendix E has a professional society that serves the technical and professional needs of engineers and students in that discipline. These societies are usually organized on both national and local levels and most support the establishment of student chapters on university campuses. The societies publish technical journals and magazines, organize technical conferences, sponsor short courses for professional development, develop codes and ethical standards, and oversee the accreditation of engineering programs in their discipline.

The benefits of getting actively involved in the student chapter corresponding to your engineering discipline will be discussed in Chapter 7. Those societies responsible for each of the engineering disciplines accredited by ABET are listed below. The discipline(s) each society is responsible for and the society website are listed. You can gain valuable information about each engineering discipline by exploring the society website.

Professional Society	**Discipline**	**Website**
American Academy of Environmental Engineers (AAEE)	Environmental Engineering	www.aaee.net
American Ceramic Society (ACerS)	Ceramic Engineering	www.ceramics.org
American Congress on Surveying and Mapping (ACSM)	Surveying and Geomatics Engineering	www.acsm.net
American Institute of Aeronautics and Astronautics (AIAA)	Aeronautical and Aerospace Engineering	www.aiaa.org
American Institute of Chemical Engineers (AIChE)	Chemical Engineering	www.aiche.org
American Nuclear Society (ANS)	Nuclear and Radiological Engineering	www.ans.org
American Society of Agricultural and Biological Engineers (ASABE)	Agricultural and Biological Engineering	www.asabe.org
American Society of Civil Engineers (ASCE)	Architectural, Civil, and Construction Engineering	www.asce.org
American Society of Mechanical Engineers (ASME)	Engineering Mechanics and Mechanical Engineering	www.asme.org
American Society for Engineering Education (ASEE)	General Engineering, Engineering Physics, and Engineering Science	www.asee.org
Biomedical Engineering Society (BMES)	Bioengineering and Biomedical Engineering	www.bmes.org
Computer Science Accreditation Board (CSAB)	Software Engineering	www.csab.org
Institute of Electrical and Electronics Engineers (IEEE)	Electrical and Electronics Engineering/ Computer Engineering	www.ieee.org

Institute of Industrial Engineers (IIE)	Industrial Engineering and Engineering Management	www.iienet2.org
Society of Fire Protection Engineers (SFPE)	Fire Protection Engineering	www.sfpe.org
Society of Manufacturing Engineers (SME)	Manufacturing Engineering	www.sme.org
Society for Mining, Metallurgy, and Exploration (SME-AIME)	Geological and Mining Engineering	www.smenet.org
Society of Naval Architects and Marine Engineers (SNAME)	Naval Architecture and Marine and Ocean Engineering	www.sname.org
Society of Petroleum Engineers (SPE)	Petroleum Engineering	www.spe.org
The Minerals, Metals & Materials Society (TMS)	Materials, Metallurgical, and Welding Engineering	www.tms.org

SUMMARY

This chapter introduced you to the engineering profession – past, present, and future. You were encouraged to take every opportunity to learn as much as you can about engineering. This will be a lifelong process, but it has already begun.

First, we helped you develop an articulate answer to a question you are likely to be asked frequently: "What is engineering?" You learned that, at its core, engineering is the process of developing a product or process that meets a customer need or perceived opportunity.

We then discussed the many rewards and opportunities that will be yours if you are successful in graduating in engineering. "Ray's Top Ten" list provided you with a picture of how an engineering degree will enhance the quality of your life – as well as the lives of others. Many of the items on the list correlate with job satisfaction – something you should place a high value on and something a career in engineering can offer you.

We then gave you a view of engineering: past, present, and future.

A view of the past came through the Greatest Engineering Achievements of the 20th Century. Reading about these not only provided an interesting retrospective of the engineering field; hopefully, it also served as an incentive to you as a new engineering student, for the achievements of the 21st century are bound to be even more spectacular than those of the 20th century. And you may be responsible for one of the "Greatest Engineering Achievements of the 21st Century." In any event, whether you look back at the past or forward towards the future, you can see what an important and exciting profession you have elected to join.

Next, we examined the present state of engineering by discussing the various academic disciplines, job functions, and employment opportunities for engineers. The North American Industry Classification System (NAICS) was used to give you a feel for the enormous economic engine that your engineering education is preparing you to be part of. We paid particular attention to the industry sectors that employ the largest numbers of engineers.

Then we discussed the future of engineering by looking at the National Academy of Engineering's "Grand Challenges for Engineering." Looking at future challenges will give you an indication of the technical fields that are expected to grow rapidly in the future. You may want to begin preparing yourself today for a career that will address one or more of those challenges.

Finally, we shined a light on the important topic of "sustainability." We pointed out the engineers will play a critical role not only in meeting the needs of the present but also doing so in such a way that the ability of future generations to meet their needs will not be compromised.

We closed the chapter by discussing engineering as a profession that you will enter when you graduate. We discussed the requirements that define a profession, including the rights and privileges that come with responsibilities and obligations. One of those rights is to become licensed as a registered Professional Engineer (P.E.).

REFERENCES

1. Constable, George and Somerville, Bob, *A Century of Innovation: Twenty Engineering Achievements That Transformed Our Lives*, Joseph Henry Press, Washington, D.C., 2003.

2. Collins, James C. and Porras, Jerry I., "A Theory of Evolution," *Audacity: The Magazine of Business Experience*, Vol. 4, No. 2, Winter, 1996.

3. Ulrich, Karl T. and Eppinger, Steven D., *Product Design and Development,* Fifth Edition, McGraw-Hill/Irwin, 2011.

4. Filippone, Antonio, "On the Possibility of Human-Powered Vertical Flight," *Journal of the American Helicopter Society*, October, 2007.

5. Larwood, Scott and Saiki, Neal, "Aerodynamic Design of the Cal Poly Da Vinci Human-Powered Helicopter," American Helicopter Society, Vertical-Lift Aircraft Design Conference, San Francisco, CA, January, 1990.

6. Naito, A., "A Study of Human-Powered Helicopter in Japan," Proceedings of the 49th Conference of the Society of allied Weight Engineers (SAWE), Paper 1925, Chandler, AZ, 1990.

7. MacCready, Paul B., "Flight on 0.33 Horsepower: The Gossamer Condor," AIAA Paper No. 78-308, Washington, DC, February, 1978.

8. Grosser, Martin, *Gossamer Odyssey: The Triumph of Human-Powered Flight*, Dover Publications, New York, NY, 1990

9. Schmaus, Joseph et al., "Design and Development of *Gamera*: A Human Powered Helicopter from the University of Maryland," Presented at the American Helicopter Society Future Vertical Lift Aircraft Design Conference, San Francisco, January, 2012.

10. Berry, Ben et al., "Design Optimization of *Gamera II*: A Human Powered Helicopter," Presented at the American Helicopter Society 68th Annual Forum, May, 2012, Fort Worth, TX.

11. Staruk, William et al., "Design and Fabrication of Ultra-Lightweight Composite Structures for the *Gamera* Human-Powered Helicopters," Presented at SAMPE 2012, Baltimore, MD, May, 2012.

12. "May 2011 National Occupational Employment and Wage Estimates - United States," U.S. Department of Labor, Bureau of Labor Statistics, www.bls.gov/oes/current/oes_nat.htm.

13. "Digest of Education Statistics 2011 Tables and Figures," U.S. Department of Education, National Center for Education Statistics. www.nces.ed.gov/programs/digest.

14. "Job Satisfaction: 2012 Edition" Report Number: TCB_R-1495-12-RR, The Conference Board, June, 2012.

15. "Building A Better Brain," *Life Magazine*, p. 62, July, 1994.

16. "NACE Salary Survey: Starting Salaries for New College Graduates," September, 2012, National Association of Colleges and Employers (www.naceweb.org).

17. "List of Billionaires (2006)," *Forbes Magazine*, 2006. www.en.wikipedia.org/wiki/List_of_billionaires.

18. Lumsdaine, Edward, Lumsdaine, Monika, and Shelnutt, J. William, *Creative Problem Solving and Engineering Design*, McGraw-Hill, New York, 1999.

19. Landels, John G., *Engineering in the Ancient World*, University of California Press, 1981.

20. "Engineering & Technology Degrees – 2011," Engineering Workforce Commission of the American Association of Engineering Societies, Inc., Reston, VA, 2011.

21. Landis, R. B., "Academic Career: It Could Be for You," American Society for Engineering Education, Washington, D.C., 1989.

22. Roberts, Edward B. and Eesley, Charles, "Entrepreneurial Impact: The Role of MIT," Kaufman Foundation, Kansas City, MO, 2009.

23. "Science and Engineering Indicators 2012," Table 3-5, National Science Foundation, SESTAT Database (sestat.nsf.gov).

24. "North American Industry Classification System (NAICS) - United States, 2002," U.S. Census Bureau, 2002, (www.census.gov/epcd/www/naics.html).

25. Friedman, Thomas L., *The World is Flat: A Brief History of the Twenty-First Century*, Farrar, Straus and Giroux, New York, 2005.

26. "The 50 Greatest Technological Inventions of the Past 25 Years," *Complex Magazine,* August 18, 2010 (www.complex.com/tech/2010/08/the-50-greatest-technological-inventions-of-the-past-25-years).

27. Beakley, G. C., Evans, D. L., and Keats, J. B., *Engineering: An Introduction to a Creative Profession,* 5th Edition, Macmillan Publishing Company, New York, NY, 1986.

28. *FE Supplied-Reference Handbook, 8th edition, 2nd revision*, National Council of Examiners for Engineering and Surveying (NCEES), November, 2010 (Available for purchase on Amazon.com or free download).

PROBLEMS

1. Review the definitions of engineering in Appendix B. Combine the best ideas from these definitions, write out your own definition, and memorize it. Ask people you come in contact with whether they know what engineering is. If they don't know, then recite your definition to them.

2. Review the National Engineers Week webpage: www.eweek.org and answer the following questions:

   a. Who are the sponsors of National Engineers Week?

   b. What is the purpose of National Engineers Week?

   c. What are some of the major activities planned for the next National Engineers Week celebration?

    d. What are some of the products available to help promote National Engineers Week?

3. Write a one-page paper about the influences (teachers, parents, TV, etc.) that led you to choose engineering as your major.

4. Pick one of the engineering guidance websites listed at the end of Section 2.1. Explore the site to learn as much about engineering as you can. Write a one-page paper summarizing what you learned.

5. Write a list of specifications for a motorized wheelchair that could be used on a sandy beach. Include performance, economic, and scheduling specifications.

6. Review the list of needed products at the end of Section 2.3. Add five additional products that you think would sell if developed.

7. Pick one of the items from Problem 6 and write a set of design specifications for the proposed product.

8. Pick one of the 20 "Greatest Engineering Achievements of the 20th Century." Write a one-page paper describing the impact of that engineering achievement on the quality of your life.

9. Create a list of activities to increase your understanding of engineering careers. Develop a plan for implementing three of these activities.

10. Add ten or more additional items to the list of rewards and opportunities of an engineering career presented in Section 2.4. Pick your top ten from the total list and rank them in order of importance.

11. Write a three-page paper on "Why I Want to Be an Engineer" by expanding on your top four items from Problem 10 and explaining why each is important to you personally.

12. Have you ever had a job you didn't like? Describe the job. What didn't you like about it? If that job played any role in your subsequent decision to major in engineering, explain what that role was.

13. Read a biography of one of the famous people listed in Section 2.4 who were educated as an engineer. Make a list of the ways that person's engineering education supported their achievements.

14. Write down five non-engineering careers (e.g., politician, entrepreneur, movie director, etc.) that you might be interested in. Discuss with a fellow student how obtaining your B.S. degree in engineering could help you pursue each of those careers.

15. What is the most challenging problem you have ever tackled in your life? Were you able to succeed at solving the problem? Did you enjoy the experience? Write a two-page essay that addresses these questions.

16. Answer the following questions related to making money:

    a. What is the legal minimum wage (per hour) in the U.S.?

    b. What is the highest hourly wage you have ever made?

    c. What hourly wage would correspond to the average starting annual salary for engineering graduates in 2012 ($60,639)?

    d. What is the hourly wage of an engineering executive making $250,000 a year? Two million dollars a year?

17. As indicated in Section 2.4, engineering graduates make up only 4.4 percent of all college graduates. Go to your career center and find out how many employers interview on campus annually. What percentage of those employers interview engineering majors only? What percentage interview business majors only? What percentage interview all other majors? What is the significance of your findings?

18. Find out how the following things work (if you don't already know):

    a. Fuel cell
    b. Radar gun
    c. Microwave oven
    d. Solar cell
    e. Digital display

    Prepare a three-minute oral presentation about one of the items that you will give at the next meeting of your *Introduction to Engineering* course.

19. Go to the National Engineers Week website "Breaking Through: The Creative Engineer" (www.eweek.org/site/nbm/intro.html). There you will find eight elements of creativity:

challenging	connecting
visualizing	collaborating
harmonizing	improvising
reorienting	synthesizing

    Pick one of these elements. Look up the definition of the term in the dictionary, study the example on the National Engineers Week webpage, and conduct further research on the element. Write a one-page paper explaining why this "element of creativity" is important in engineering work.

20. Go to Professor John Lienhard's *The Engines of Our Ingenuity* webpage: www.uh.edu/engines. Pick three of the more than 2,800 episodes you will find there. Study those three. Write a two-page paper on why you picked the ones you did and what you learned from studying them.

21. In Section 2.6 and Appendix E you learned that engineering disciplines can be divided into two categories: (1) the eight largest disciplines (electrical, mechanical, civil, computer, chemical, biomedical, industrial, and aerospace), (2) a larger number of smaller, more specialized disciplines. Make a list of the advantages and disadvantages of selecting your major in one or the other of these two categories.

22. Which of the engineering disciplines listed in Section 2.6 and described in Appendix E are offered by your university? Find out how many students graduate annually from your institution in each of these disciplines.

23. Pick one of the engineering disciplines listed in Section 2.6 and described in Appendix E. Visit the webpage of the professional society corresponding to that discipline and note any information that applies specifically to engineering students. Share what you learned with your classmates in your next *Introduction to Engineering* class.

24. Pick one of the engineering disciplines listed in Section 2.6 and described in Appendix E. Write a three-page paper that would serve to inform a high school senior about that discipline.

25. Pick one of the technical divisions or societies of either the American Society of Mechanical Engineers (ASME), the Institute of Electrical and Electronics Engineers (IEEE), or the

American Society of Civil Engineers (ASCE) listed in Appendix E that you would like to know more about. Research the division or society and write a one-page paper describing it.

26. Which of the civil engineering specialties described in Appendix E would provide you the greatest opportunity to benefit society? In what ways?

27. Go to the U.S. Bureau of Labor Statistics' web-based "Occupational Outlook Handbook" at: www.bls.gov/oco/ocos027.htm. Study the information there to learn as much as you can about "engineers." What does it say about the job outlook for engineers?

28. Go to the American Institute of Chemical Engineers (AIChE) "CareerEngineer" webpage: www.aiche.org/resources/careers. Click on "Find a Job," and read about the ones listed there. Which one appeals to you the most? Prepare a two-minute talk describing its appeal to your *Introduction to Engineering* classmates.

29. Interview a practicing engineer. Find out answers to the following questions:
    a. What engineering discipline did he or she graduate in?
    b. To what extent do the knowledge and principles of that discipline apply in his or her current job?
    c. What industry sector does he or she work in?
    d. What percentage of his or her time is spent in the various engineering job functions (design, test, development, management, etc.)?

30. Develop a list of attributes that would be desirable for each of the engineering job functions described in Section 2.7. Which of these job functions appeals to you? Be ready to explain your reasons in a discussion at the next meeting of your *Introduction to Engineering* class.

31. Familiarize yourself with the 2012 NAICS system by doing the following exercise. Begin by accessing the Internet. Then proceed as directed below:
    a. Go to: www.census.gov/naics.
    b. Under "2012 NAICS Search," enter "334."
    c. Browse through all the categories of products listed under NAICS 3345.
    d. Find the products listed under NAICS 334510 and print them out.

32. Pick one of the products listed under NAICS 334510 from Problem 31 and research what companies manufacture that product. Contact that company and investigate how they use engineers in the design, manufacturing, and testing of that product, as well as in the marketing of it. Write a summary of what you learned.

33. Learn about the "Engineering Services" industries by following the steps outlined in Problem 31 and entering the six digit code "541330." How many entries did you find? Would you be interested in working in this industry? Why or why not?

34. Obtain a list of employers that conduct on-campus interviews of engineering graduates through your career center. Try to identify which industry sector each employer belongs in, based on those listed in Section 2.8. Do some of the employers fit into more than one industry sector?

35. Identify the two or three engineering disciplines that you think would be most closely associated with each of the eight "Service" Economic Sectors and six "Manufacturing" Economic Subsectors described in Section 2.8.

36. Make a list of ten products that would be manufactured by each of the six "Manufacturing" Economic Subsectors listed in Section 2.8.

37. Pick one of the "50 Greatest Technological Inventions of the Past 25 Years." Write a short paper discussing how the invention has improved the quality of your life. Also discuss the ways in which the invention might interfere with your goal of graduating in engineering. Develop a plan to minimize any negative impact the invention might have on your education.

38. Explain how each of the "Grand Challenges for Engineering" listed in Section 2.9 will impact your future. What effect will each have on the engineering job market? (Will it increase or decrease the number of jobs? In which disciplines? Will it change the nature of current jobs?)

39. Pick one of the ten "important attributes of green engineering and sustainable designs" listed on Page 77. Research the attribute and write a two-page paper explaining:
    a. What is meant by the attribute?
    b. Why is it important?
    c. What are some strategies for accomplishing it?
    d. What are some of the difficulties in accomplishing it?

40. Make a list of (1) the rights and privileges and (2) the responsibilities and obligations you will have when you join the engineering profession.

41. Obtain information about the process of becoming a registered Professional Engineer in your state. How do the requirements and procedures differ from those presented in Section 2.10? What engineering disciplines are licensed in your state? Set a personal goal of passing the FE Exam before or soon after you graduate. Develop a set of strategies that will ensure you are well prepared to pass the exam.

CHAPTER 3
Understanding the
Teaching/Learning Process

Education is not preparation for life; education is life itself.

— John Dewey

INTRODUCTION

In Chapter 1 we identified "approach" as a key factor that will lead you to success in your engineering studies. We linked "approach" with "effort," another key factor, explaining that the successful student is one who works both *hard* and *smart*. In this and the two subsequent chapters, we will focus on what it means to work "smart."

This chapter provides an overview of the teaching/learning process. Understanding this process will help you take full advantage of both the teaching part and the learning part.

We begin by discussing the learning process. We define learning and describe its components – receiving and processing new knowledge, and demonstrating mastery of that knowledge. You will have the opportunity to find out how you prefer to receive and process new knowledge – insights that will be useful to you in designing your own learning process. We will also describe the characteristics of "expert" learners and encourage you to continuously improve your learning skills by observing your learning process and making changes based on those observations.

We conclude the section on learning with the important perspective that learning is a reinforcement process. This is an overarching principle that you should take advantage of at every opportunity.

Next, we provide a brief overview of the teaching component of the teaching/learning process. We point out the ways teaching is delivered, including various teaching styles used by your professors. Being aware of how you are taught can enhance your learning experience.

Then we point out some common mistakes students make as they transition from high school to college-level engineering study. A serious (and costly) mistake is assuming that the same strategies and approaches that worked in high school will work here. We also provide an indication of approaches and strategies for avoiding these mistakes.

We conclude the chapter with a perspective on the subject of seeking help. Failing to utilize available resources – particularly professors and fellow students – is another mistake that some students make. "Standing on the shoulders of others" is fundamental to the very concept of an education.

3.1 WHAT IS LEARNING?

Although you've been an active participant in the learning process for many years, it is unlikely you have had any formal training in it. Becoming an expert learner requires not only that you devote time and energy to learning, but that you devote time and energy into learning *how* to learn.

Learning, broadly defined, is the process of acquiring:

- New knowledge and intellectual skills (cognitive learning)

- New manual or physical skills (psychomotor learning)

- New emotional responses, attitudes, and values (affective learning)

COGNITIVE LEARNING

Cognitive learning is demonstrated by knowledge recall and higher-level intellectual skills. Since acquiring new knowledge and intellectual skills will be the primary focus of your engineering education, it's useful to explore the process in some detail.

Bloom's "Taxonomy of Educational Objectives" [1] identifies six levels of intellectual skills within the cognitive domain. At the lowest level is the simple recall or recognition of facts. At the highest level is creativity. In between are increasingly more complex and abstract mental levels. The following are brief definitions of each level of intellectual skill, along with examples that represent the activities for each level.

Intellectual Skill	Definition	Examples of Intellectual Activity
Remembering	Retrieving, recognizing, and recalling relevant knowledge from long-term memory	arranging, defining (from memory), duplicating, labeling, listing, memorizing, naming, ordering, recognizing, relating, recalling, repeating, reproducing, stating
Understanding	Constructing meaning from oral, written, and graphic messages	classifying, describing, defining (in your own words), discussing, explaining, expressing, identifying, indicating, locating, recognizing, reporting, restating, reviewing, selecting, translating
Applying	Carrying out or using a procedure through executing or implementing	applying, choosing, demonstrating, dramatizing, employing, illustrating, interpreting, operating, practicing, scheduling, sketching, solving, using, writing
Analyzing	Breaking material into constituent parts; determining how the parts relate to one another and to an overall structure or purpose	analyzing, appraising, deriving, calculating, categorizing, comparing, contrasting, criticizing, differentiating, discriminating, distinguishing, examining, experimenting, predicting, questioning, testing

Evaluating	Making judgments based on criteria and standards through checking and critiquing	appraising, arguing, assessing, attaching, choosing, comparing, contrasting, critiquing, defending, judging, predicting, rating, selecting, supporting, valuing, evaluating
Creating	Putting elements together to form a coherent or functional whole; reorganizing elements into a new pattern or structure	arranging, assembling, collecting, composing, constructing, creating, designing, developing, formulating, managing, organizing, planning, preparing, proposing, setting up, writing

In high school you primarily worked at the first two levels of demonstrating subject mastery – remembering and understanding. Often you could earn good grades by memorizing material and successfully repeating it back on exams or other assignments. In your university education, the expectations are higher. From the start, you will be expected to think at levels 3 and 4 – applying and analyzing – mastering concepts and applying them to solve new problems. Your performance will be evaluated on your ability to apply what you learn in one context to new contexts. In your first two years you may be given some problems that require you to think at levels 5 and 6 – evaluating and creating, but the strongest focus on these levels of thinking will come in your junior and senior years.

The importance of acquiring higher intellectual skills is underscored in an excellent book by John Bransford titled *How People Learn* [2]:

> All learning involves "transfer" – defined as the ability to extend what has been learned in one context to new contexts – from previous experiences. Educators hope that students will transfer learning from one problem to another within a course, from one school year to another, between school and home, and from school to the workplace. Transfer is affected by the degree to which people learn with understanding rather than merely memorize sets of facts or follow a fixed set of procedures.

I can assure you that mastering the learning skills presented in this chapter and the next two chapters will aid you in this process of "kicking it up a notch" from what was expected of you in high school.

PSYCHOMOTOR LEARNING

Psychomotor learning is demonstrated by physical skills – coordination, dexterity, manipulation, grace, strength, and speed, for example, actions that demonstrate the fine motor skills such as use of precision instruments or tools, or actions that demonstrate motor skills such as dance or athletic performance.

Within engineering education, examples of learning in this domain might include sketching, computer keyboard skills, machine tool operation, and certain laboratory skills.

AFFECTIVE LEARNING

Affective learning is demonstrated by behaviors indicating attitudes of awareness, interest, attention, concern, and responsibility. This domain relates to emotions, attitudes, appreciations, and values, such as enjoying, respecting, and supporting. The ability to listen and respond in interactions with others is part of affective learning.

Many of the topics in this text fit into the affective learning category. Indeed, one of the primary purposes of this book is to guide your personal development in many of the areas listed above. This development will continue throughout your engineering education, even though it might not be addressed explicitly in your required coursework. For example, through your participation in study groups and laboratory and design project groups, you will grow in your ability to "listen and respond in interactions with others." However, you might have to elect to take a course in Interpersonal Communication to gain explicit training in this area. By doing so, you will avoid being the person in the cartoon below.

© Randy Glasbergen / glasbergen.com

GLASBERGEN

"I'm trying to be a good listener, but you
keep breaking my concentration by talking!"

REFLECTION

Reflect on the three categories of learning – cognitive learning, psychomotor learning, and affective learning. What are some examples of learning you have experienced in each of the three categories? Undoubtedly you have acquired new knowledge (cognitive learning) and developed dexterity in some area such as typing or sketching or dancing (psychomotor learning). Do you have more difficulty coming up with examples under the affective learning category? Are you open to learning in areas such as enjoying, respecting, and supporting others?

3.2 HOW DO WE LEARN?

The process of learning new knowledge and intellectual skills (cognitive learning) involves two steps: 1) receiving new knowledge, and 2) processing that new knowledge.

RECEIVING NEW KNOWLEDGE

There are two key aspects to the way you receive new knowledge:

- The type of information you prefer
- The sensory channel through which you most effectively perceive external information

Research has shown that learners' preferences differ in each of these aspects.

I. What Type of Learner Are You?

Sensing learners. Sensing learners focus on things that can be seen, heard, or touched. They like facts and data, the real world, and above all, relevance. They are patient with details and enjoy solving problems by well-established methods.

Intuitive learners. Intuitive learners are dreamers. They prefer ideas, possibilities, theories, and abstractions. They look for meanings, prefer variety, and dislike repetition. They tend to dislike "plug and chug" courses and are impatient with detailed work.

II. Through What Sensory Channel Do You Perceive External Information Most Effectively?

Visual learners. Visual learners learn more effectively through the use of pictures, diagrams, flowcharts, graphs, sketches, films, and demonstrations.

Verbal learners. Verbal learners respond more to the written or spoken word. They like to read about things or hear explanations from an expert.

REFLECTION

When you receive new knowledge, what type of information do you prefer? Are you a sensing learner? Or are you an intuitive learner? Do prefer to receive information visually? Or verbally? What implications do your preferences have for your learning process?

PROCESSING NEW KNOWLEDGE

Just as there are two ways to receive new knowledge, there are two aspects to processing that knowledge:

- the way you prefer to process information
- the way you progress toward understanding

Research has shown that learners differ with regard to their preferences in these two areas.

III. The Way You Prefer to Process New Knowledge

Active learners. Active learners tend to process information while doing something active with it. Consequently, active learners think out loud, try things out, and prefer group work. Sitting through lectures is particularly hard for active learners.

Reflective learners. Reflective learners prefer to think about information quietly first. They want to understand or think things through before attempting to do anything themselves. They tend to prefer working alone.

IV. The Way You Progress Toward Understanding

Sequential learners. Sequential learners prefer linear steps, with each step following logically from the previous one. They tend to follow logical stepwise paths in finding solutions.

Global learners. Global learners tend to learn in large jumps, absorbing material almost randomly without seeing connections, and then suddenly "getting it." They prefer to see the "big picture" and then fill in the details.

REFLECTION

How do you prefer to process new information? Are you an active learner who wants to do something with it? Or are you a reflective learner who wants to think about it first? How do you prefer to progress toward understanding? Are you a sequential learner, preferring to take logical steps? Or are you a global learner, preferring to start with the big picture and then fill in the steps? Is there any value to understanding how you prefer to process new information?

INDEX OF LEARNING STYLES QUESTIONNAIRE

You probably already have a sense of your preferred learning styles, but have never thought to articulate them. If you'd like to get an even more definitive indication, I encourage you to complete the *Index of Learning Styles Questionnaire* developed by Barbara Soloman and Richard Felder at North Carolina State University. The questionnaire can be completed online at:

Richard Felder

www.engr.ncsu.edu/learningstyles/ilsweb.html

At that website, you will be asked to choose one of two preferences for 44 items that cover four areas (two areas related to receiving information and two areas related to processing information).

You will immediately receive scored results indicating whether you have a strong preference, moderate preference, or are fairly well-balanced on each of the four scales.

As you'll learn later in this chapter, if your preferred ways of receiving new knowledge are *verbal* and *intuitive* and your preferred ways of processing new knowledge are *sequential* and *reflective*, you are in step with the ways you are most likely being taught (i.e., traditional lecture format). However, you can almost certainly benefit by adopting the full spectrum of learning styles in your studies (sensory as well as intuitive; visual as well as verbal; active as well as reflective; global as well as sequential).

If your preferred ways of receiving new knowledge are *visual* and *sensory* and your preferred ways of processing new knowledge are *global* and *reflective*, you may find a mismatch between the way you prefer to learn and the way you are being taught. The good news is that you are primarily responsible for creating your learning experience. Just make sure the way you study is compatible with your preferred learning styles. Some general advice from Richard Felder as to how to do this can be found at: www.ncsu.edu/felder-public/ILSdir/styles.htm .

Keep in mind, however, that:

All learners benefit from using learning styles on both sides of all four dimensions.

Doing things that are compatible with your style can help you compensate for mismatches with the dominant style of your teachers. Doing things that are on the opposite side of your learning style preference will give you a perspective on the material you might not normally get, while helping you develop skills that will enhance your professional success – skills you might not develop if you only followed your natural inclinations.

3.3 METACOGNITION – THE KEY TO IMPROVING YOUR LEARNING PROCESS

Famous Swiss psychologist Carl Jung put forth the concept of the "observing ego." This is the part of you that observes your "self." It observes what you do. It observes how you think. And it observes how you feel. And through this process of observing, it feeds back information (assessment) to enable you to make changes. Some examples of this process are shown below.

Activity	Observation	Feedback	Change
Running a 10K race	I'm running too fast.	I'd better slow down.	Slow down pace.
Falling asleep while studying	I'm not going to get my homework done.	A break will help.	Take a walk around the block.
Doing homework problems	I can't do Problem #5.	I need to seek help.	Go see instructor during office hours.
Attending math class	I'm not following what's being presented.	I could get clarification by asking a question.	Raise hand and ask question.

Improving your learning process by observing it, developing feedback on what you observe, and making changes based on that feedback are all part of *metacognition*. When engaging in the metacognition process, you should follow these steps:

1) Plan your learning
2) Monitor your learning
3) Evaluate your learning

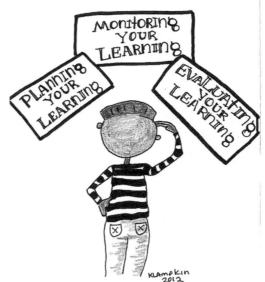

I would encourage you to begin regularly observing your learning process. The easiest way to do this is to ask yourself questions.

1. When ***planning your learning,*** ask questions like:

- What in my prior knowledge or experience will help me with this particular task?
- What should I do first?
- Why am I doing this task?
- How much time do I have to complete this task?

2. When ***monitoring your learning***, ask questions like:

- How am I doing?
- Am I on the right track?
- How should I proceed?
- What information is important to remember?
- Should I move in a different direction?
- Should I adjust the pace based on the difficulty of a subject (too difficult – slow down; too easy – speed up)?
- What do I need to do if I do not understand?

3. When **_evaluating your learning,_** ask questions like:

- How well did I do on a particular task or assignment?
- Did my particular course of action produce more or less than I had expected?
- What could I have done differently?
- How might I apply this line of thinking to other problems?
- Do I need to go back through the task to fill in any blanks in my understanding?

REFLECTION

Think about the following characteristics of "expert" learners:

- Control the learning process rather than become a victim of it
- Are active, not passive, in their approach to learning
- Are motivated (e.g., enjoy learning, have short-term and long-term goals, etc.)
- Are disciplined (i.e., have learned good habits and use them consistently)
- Are more aware of themselves as learners (e.g., know their own strengths and weaknesses)
- Initiate opportunities to learn
- Set specific learning goals for themselves
- Have a larger repertoire of learning strategies from which to choose
- Know not only *what* to learn, but *how* to learn
- Plan their approach to learning
- Monitor their learning while it's happening (e.g., notice when they're not learning and adjust their learning approach)
- Are more adaptive because they *do* self-monitor while learning
- Reflect more upon their own learning
- Evaluate the effectiveness of learning approaches and strategies
- Are more sensitive to the demands of specific academic tasks
- Use learning strategies selectively
- Tend to attribute failures to correctable causes
- Tend to attribute successes to personal competence

Rate yourself on a scale of zero to ten on each of these items. Are there areas that need improvement? Are you willing to work at improving? Pick two or three areas you would like to improve in, and for each come up with three specific things you can do for each.

I hope I have persuaded you of the benefits of metacognition – that is, "observing" your learning process with the goal of becoming an expert learner.

3.4 LEARNING IS A REINFORCEMENT PROCESS

There is one more important principle that should influence every aspect of your participation in the teaching/learning process. It is that:

> ### *Learning is a reinforcement process.*

A critical part of the learning process is what we call "reinforcement." That is, learning comes from repeated exposure to subject material – the more the better. Consider, as an example, the way we master the subject of "mechanics."

An Example: The Study of Mechanics

The way in which we learn the subject of mechanics, the study of forces and motion, can illustrate the importance of reinforcement in the learning process.

Our first exposure to mechanics may have come in high school physics. Next, we study a whole semester of mechanics in our freshman physics course. In our sophomore year, we may have a course in statics and, in our junior year, a course in engineering dynamics. If we are interested, we can take several senior-level courses and, for a thorough understanding of mechanics, we could pursue graduate study – a Master's or even Ph.D. degree.

Even then, if we were to begin to teach mechanics, we would find areas in which we were not completely clear, and probably only after a number of years of teaching would we feel that we were even approaching total mastery of the subject.

The point of this example is not to discourage you, but rather to encourage you to take advantage of every opportunity to reinforce your learning. Even for the brightest person, learning is a slow process that occurs over time and relies on repeated reinforcement. By knowing this, hopefully you will not fall into the common trap of thinking that you can "cram" in the material the night before a test (like the student in the picture).

The educational system is structured to give you the opportunity to reinforce the subject matter many times within a semester or quarter. Here are examples of the reinforcement process at its best.

When	What To Do
Before class	Prepare for the lecture by reviewing notes, reading text, attempting a few problems, formulating some questions
During class	Attend lecture, concentrate intently, take detailed notes, ask questions
After class, but before next class meeting	Review and annotate notes, reread text, work assigned problems, work extra problems, meet with a study partner or study group to go over material and problems
In preparation for test or exam	Review notes, review text, rework problems, meet with a study partner or study group to go over material and problems
In preparation for final exam	Review notes, reread text, rework problems, meet with a study partner or study group to go over material and problems

This systematic approach to learning that involves repetition, review, and reinforcement will carry you a long way toward becoming an expert learner. Once you master the specific skills presented in the next two chapters as well, you'll be an expert learner for sure.

If you keep the principle that "learning is a reinforcement process" in mind as you design your learning process, you won't find yourself in the situation in this cartoon.

Copyright 2006 by Randy Glasbergen.
www.glasbergen.com

**"I forgot to make a back-up copy of my brain,
so everything I learned last semester was lost."**

3.5 UNDERSTANDING THE TEACHING PART OF THE TEACHING/ LEARNING PROCESS

The teaching part of the teaching/learning process is primarily achieved by the following well-known *teaching modes*:

- Large lectures, in which one professor teaches 100-300 or more students

- Small lectures, in which one professor teaches 20-30 students

- Recitations, in which a teaching assistant reviews the material and solves problems for small groups of ten to 15 students

- One-on-one tutoring, in which one tutor works with one student

Despite their obvious differences, all four teaching modes have one feature in common. Each involves a knowledgeable person communicating what he or she knows to a less knowledgeable person. Generally, most of the communication is one-way – i.e., from the teacher, teaching assistant, or tutor to the student. And most importantly, **_students learn relatively little_** from participating in any of these modes.

That last statement should alarm you or at least cause you to question how I could make such a provocative claim. Here's how. Imagine you're in an engineering course and your professor introduces a new principle. You go to the lecture, recitation, and tutoring sessions, but you don't do anything outside of those activities. Then you are given an exam on the principle. What score would you expect to make?

The limited effect of these teaching modes – especially the large lecture format – becomes quite apparent if you envision the process as one educator has aptly described it:

> *The information passes from the notes of the professor to the notes of the student without passing through the mind of either one.*

That there is a major difference between high school and university-level study is illustrated by the 80/20 rule [3]. In high school roughly 80 percent of what you needed to know came from the teacher and in-class work. Only 20 percent of the learning occurred outside the classroom. In college, this rule is reversed. Only 20 percent of what you learn will come from the professor and class lectures. Eighty percent of what you learn will come from work you do outside the classroom. *This is perhaps the most important concept you must grasp about the college learning environment.*

TEACHING STYLES

Even within the general lecture format, there are variations in the way teaching is done. In reading Section 3.2, you should have discovered your preferred learning styles. Your effectiveness in creating your learning experience will be determined not only by your understanding of the ways you best learn, but also by seeing them in the context of how you are being taught.

The following is an overview of the various dimensions of teaching styles, outlined by Dr. Richard Felder of North Carolina State University (and author of the Foreword to this book), a world-renowned expert on teaching and learning styles in engineering education [4]. Dr. Felder examines five categories of the teaching process and describes two possibilities for the approach your professors might take for each of these categories. The teaching styles most prevalent in math/science/engineering courses (abstract, verbal, deductive, passive, and sequential) are underlined.

1) **What type of information is emphasized?**

   Concrete – Facts, data, observable phenomena
   Abstract – Principles, concepts, theories, mathematical models

2) **What mode of presentation is stressed?**

   Visual – Pictures, diagrams, films, demonstrations
   Verbal – Spoken words, written words

3) **How is the presentation organized?**

   Deductive – Start with fundamentals, proceed to applications
   Inductive – Start with applications, proceed to fundamentals

4) **What mode of student participation is facilitated?**

   Active – Student involved (talking, moving, reflecting, solving problems)
   Passive – Student as spectator (watching, listening)

5) **What type of perspective is provided on the information presented?**

   Sequential – Step by step progression
   Global – Context and relevance are provided

REFLECTION

Think about the classes you are taking in terms of each of the five categories listed on the previous page. Do your instructors predominantly use the teaching styles that are underlined or do they use the other one? Do they use some combination of the two? Pick the teaching style under each of the five categories that you would learn best from and place an asterisk next to it. How many match the ones most common in your courses? How many differ? What meaning does this have for you?

By now, I hope you're asking yourself some very good questions, like:

- What value is it to me to understand how my professors teach?

- If the way my professors teach differs from the way I prefer to learn, does that mean I can't learn the material and I am perhaps in the wrong major?

- If some students prefer to be taught one way and other students prefer to be taught another way, why wouldn't the professor use both ways?

These are complex and interesting questions and I would encourage you to seek out the answers. Here are some of my thoughts for starters.

BENEFITS OF UNDERSTANDING HOW PROFESSORS TEACH. There are several benefits to understanding the different ways of teaching. Perhaps primary among these is that the knowledge will guide you in designing your learning process.

But equally important is that you will be doing lots of teaching yourself. When I was Dean of Engineering, someone would ask me from time to time, "Do you do any teaching?" Even though I knew they really meant "Was I teaching any courses?" and I wasn't, I would respond, "I'm always teaching." Throughout your time as a student and during your professional career, you will constantly be teaching others what you know – in formal and informal meetings, through your written communications, and in almost everything you do.

Understanding the different ways of transmitting knowledge will be an extremely useful skill in all aspects of your life and career. And who knows? Maybe someday you'll decide to become an engineering professor.

WHAT IF THE WAY I PREFER TO LEARN DIFFERS FROM THE WAY I AM TAUGHT? Rather than view this as a problem, I suggest you view it as an opportunity. Just because you prefer to be taught one of two possible ways doesn't mean you can't learn the other way. During your learning process, you can learn even more by translating what you were taught into the way you prefer to learn. For example, if you need to know the context and relevance (global perspective) for what you are being taught and your professor doesn't provide it, it's likely that you will learn even more if you develop it on your own than if the professor had provided it.

Also remember that there is a difference between preference and competence. You may like doing something but not be good at it. For example,

I love to sing but I can't carry a tune. Conversely, you may be good at something, but not like doing it. I'm sure I'd be good at accounting, but I wouldn't want to do it for a living.

Strive to improve your ability to get the most out of all teaching styles – the ones you prefer and the ones you don't. You might prefer to study by yourself but find that you are very good studying collaboratively with other students. As a student and as a professional, you'll be learning from and teaching people who have different preferences from yours. Your effectiveness will depend on your ability to use different ways of getting your points across.

The bottom line is you will benefit by developing your skills on both sides of the four learning style dimensions. Good engineers are observant, methodical, practical, and willing to check calculations and replicate experiments over and over to be sure they're right (sensing skills). They also have to be innovative, deal with theories and models, and think deeply about the meaning of their observations and results (intuitive skills). Good engineers have to deal with both visual and verbal information. They must both reflect on things and also take action. And they need to both be aware of the big picture and also proceed in a stepwise manner.

REFLECTION

Think about your ability to do various things. Are there things you like to do, but don't do well? Are there things that you do very well, but don't like doing? Or do you generally only like doing the things you do well? How can understanding the difference between preferences and competence benefit you in your education?

Above all, there is no need for you to drop out of school or to change majors. You can learn

to learn no matter how you are being taught. Besides, there's a place in the engineering profession for individuals with different preferences for the way they are taught. Think about the various engineering job functions described in Chapter 2 – analysis, design, test, development, sales, research, management, consulting, and teaching. Although one individual might be best suited for one job function (e.g., someone who prefers to be taught visually might make a good design engineer) and another individual might be best suited for another job function (e.g., someone who prefers to be taught verbally might make a good analytical engineer), all engineering job functions require individuals who can both learn and teach in a variety of styles.

WHY DON'T YOUR PROFESSORS USE A VARIETY OF TEACHING STYLES? Many do. And in time, more and more will. There is an impetus for change within engineering education. *Interactive lectures, problem-based learning, inquiry-guided learning,* and *just-in-time teaching* are examples of teaching methods that are gaining acceptance within engineering education. These methods bring more student involvement, context, and relevance into the classroom.

But change is slow. Formal training in teaching methods is not a required part of the process of becoming a math, science, or engineering professor. In the absence of such training, most professors tend to teach the way they were taught. And so the most prevalent teaching styles of the past (abstract, verbal, deductive, passive, sequential) tend to be propagated into the future.

Whatever you do, don't use the way your professor teaches as an excuse for not learning. If you believe you are having difficulty because of the way you are being taught, you might speak to your professor and suggest ways in which he or she could help you get more out of the lectures. You might ask, for example, that the professor work out more problems in class. Or if you tend to be a global learner, ask the professor to address "the big picture." You might also ask your professor to recommend additional resources that would aid you in your learning process. One caution: Do this politely and in a constructive tone. No one likes to think he or she is being criticized.

3.6 MISTAKES STUDENTS MAKE

As we discussed in Chapter 1, engineering study is challenging, but you have the ability to do it. What's critical for your success is to avoid many of the mistakes that students make early on. If you do, you're likely to be one of the many who succeed. If you don't, you'll increase the chance you'll be among those who don't succeed.

A big mistake is to assume your college engineering studies will be like high school and the same strategies and approaches that worked there will work here. In reality, the faster pace and higher expectations for learning require new strategies and improved learning skills. Included with each mistake listed below is a general strategy for overcoming it. Taking advantage of these success strategies will require that you absorb, practice, and refine the learning skills that are described in this chapter and in Chapters 4 and 5.

Mistakes Students Make	Strategies for Overcoming Them
Assume engineering study will be like high school.	Work to understand and adjust to the differences between high school and college-level engineering study.
Program yourself for failure through too many commitments.	Create a life situation that enables you to devote adequate time and energy to your studies.
Spend little time on campus.	Immerse yourself in the academic environment of the institution.
Neglect studying.	Schedule study time. Devote significant time and energy to studying.
Delay studying until test is announced.	Master the material presented in each class prior to next class.
Study 100% alone.	Study collaboratively with other students.
Come to each lecture unprepared.	Review notes, read text, and attempt problems prior to each lecture.
Avoid professors.	Interact regularly with professors outside the classroom.
Cut classes and/or don't get the most out of lectures.	Attend classes and practice good listening skills. Ask questions in class.

Fail to take notes or take notes but fail to use the notes properly in the learning process.	Take effective notes and use a systematic learning methodology to study from notes.
Skim over the material in an assigned chapter in a rush to get to the assigned homework problems.	Use reading for comprehension methodology (see Section 5.1) to understand the general concepts thoroughly before attempting problems.
Fail to solve assigned problems. Don't approach problems using a systematic problem solving method.	Solve not only assigned problems but extra problems; use systematic problem solving methods.

Our purpose in this chapter and in Chapters 4 and 5 will be to guide you in developing important learning strategies and skills that will move you from the left-hand column into the right-hand column. If you are able to do this, the likelihood you will succeed in your engineering studies will be greatly enhanced.

REFLECTION

Review the list of "Mistakes Students Make." Do any of these describe the way you are approaching your engineering studies? Do you think you would benefit by changing to the strategy described in the right-hand column? Pick one or two items and make a commitment to make the required changes.

3.7 DON'T BE HUNG UP ON THE IDEA OF SEEKING HELP

Do you feel that seeking help is a sign of weakness? That if you make it on your own you will get more out of your education? The idea that we can make it through life without help from others is simply not true.

The truth is we all rely heavily on others to live, grow, and thrive. We come into this world totally dependent on our parents or guardians for our very survival. As we grow, most of what we learn we are taught by others – parents, family, teachers, peers. In school when we use a textbook, in engineering or any other discipline, we benefit from the many experts who have evolved the subject over years to the point where we can readily understand it. This point is perhaps best underscored by a famous quote by Isaac Newton, who is generally viewed as one of the greatest thinkers in the history of humankind:

> ***If I have seen further, it is by standing on the shoulders of giants.***

These few observations alone should be sufficient to disabuse you of the notion that you can go it alone. Although part of our country's early mythology glorified the rugged individualist, this image was always seen as a romantic ideal – nice to dream about but impossible to become.

If you have somehow been led to believe that working independently is the best way to approach your engineering studies, think again. Don't let such misconceptions stand in your way – as they undoubtedly will. Instead, take full advantage of the many resources and learning opportunities your campus offers you. The really smart student does.

At your college or university, there are two immediate resources available to help you with your academic work:

➤ **Your peers**

➤ **Your professors**

The value of making use of your peers and your professors is best explained in the following excerpt from an excellent Harvard University study on the teaching/learning process [5]:

> Is there any common theme that faculty members can use to help students, and indeed that students can use to help themselves? The answer is a strong yes. All the specific findings point to, and illustrate, one main idea. It is that students who get the most out of college, who grow the most academically, and who are the happiest, organize their time to include interpersonal activities with faculty members, or with fellow students, built around substantive academic work.

I assume that you want to be one of those students who get the most out of college, who grow the most academically, and who are the happiest. Now you know what you need to do: Organize your time to include interpersonal activities with faculty members, or with fellow students, built around substantive academic work. However, you may not know <u>how</u> to bring about these results. In Chapter 4 we will provide you with specific strategies on making effective use of your professors and will discuss other sources of help available to you including tutoring and other academic resources. In Chapter 5 we will address the important issue of making effective use of your peers.

REFLECTION

Are you a person who likes to help others? Do you like others to help you? If you get pleasure from helping others, why wouldn't you want to give others the same pleasure by letting them help you? Imagine that even after giving it your best effort, you just can't understand some material that was presented in a class. Who would you be most inclined to ask for help? Your professor? A classmate? No one?

Many online sites also can provide various types of help. One excellent resource that can help you make up for any deficiencies you might have in mathematics and science is the Khan Academy website (www.khanacademy.org). There you will find 3,200 videos, each approximately ten minutes in length, covering mathematics from arithmetic to calculus and science from biology to chemistry and physics. I hope you'll check out a few of these videos to see what is there and whether any can help you.

There are many online sites that can provide you with knowledge and help develop your study skills. Two of the best are the "Study Guides and Strategies" website: www.studygs.net and the "College Survival" website: www.universitysurvival.com/student-success-skills of the University of West Virginia. On these two websites you will find lots of information on topics related to studying and learning, time management, reading, writing, and thinking. Mastering much of the material there would go a long way toward developing you into a "master" student.

3.8 ACADEMIC SUCCESS SKILLS SURVEY

Before you go to the next chapter, complete the *Academic Success Skills Survey* at the end of this chapter. Note any items you check as "Neutral," "Disagree," or "Strongly Disagree" and commit to improving yourself in those areas. As you study the subsequent chapters of this text, give particular attention to the sections that address those areas.

SUMMARY

This chapter provided an introduction to the teaching/learning process. You may not have previously given much thought to how this process works. Hopefully you now understand that the institution focuses primarily on the *teaching* part, while the *learning* part is left up to you.

We began by looking at the learning process. We defined learning and described its component parts. You identified your preferred ways of receiving and processing new knowledge – an awareness that can aid you in designing your learning process. We also encouraged you to continuously refine your learning process through metacognition: closely observing your learning process, feeding back what you observe, and making changes based on your observations.

We also provided you with a perspective on the importance of reinforcement in the learning process. Taking a systematic approach to your learning that involves repetition, review, and reinforcement will go a long way toward making you an expert learner.

Then we turned to the teaching part of the teaching/learning process. We discussed the various teaching styles used by your professors and the benefits of understanding how your professors teach.

Next we pointed out mistakes commonly made by students as they transition from high school to university-level engineering study. Your success will depend to a great extent on your ability to avoid these mistakes.

We closed the chapter with a perspective on the concept of seeking help. I hope you are persuaded that "standing on the shoulders" of others is fundamental to the very concept of an education and you will take full advantage of all the resources available to you – particularly your peers and your professors.

In the next chapter, we will focus on what you can do to get the most out of the teaching part of the teaching/learning process. By practicing the approaches presented there, you will lay the strongest possible foundation on which to build your learning process.

REFERENCES

1. Anderson, L. W. and Krathwohl, D. R. (editors), *A Taxonomy for Learning, Teaching, and Assessing: A Revision of Bloom's Taxonomy of Educational Objectives, Complete Edition*, Longman, New York, NY, 2001.

2. Bransford, John, Brown, Ann L., and Cocking, Rodney R., *How People Learn: Brain, Mind, Experience, and School: Expanded Edition*, National Academies Press, 2000.

3. "Why 'Good' Students Do Bad in College: Impactful Insights," *The Well*, (www.thewelledu.com), June 4, 2012.

4. Felder, Richard M. and Silverman, Linda K., "Learning and Teaching Styles in Engineering Education," *Engineering Education*, v. 78(7), pp. 674-681, 1988.

5. Light, Richard J., *The Harvard Assessment Seminars: Second Report*, Harvard University, Cambridge, MA, 1992.

PROBLEMS

1. Review the 11 attributes presented in Chapter 1 that the ABET *Engineering Criteria 2000* mandates for engineering graduates. Which primarily involve cognitive learning? Which are most likely to involve psychomotor learning? What psychomotor skills might you acquire? Which of the attributes are most likely to involve affective learning? Which affective skills might you acquire?

2. Consider the 11 attributes presented in Chapter 1 that the ABET *Engineering Criteria 2000* mandate for engineering graduates. In the table below indicate the extent to which each requires one of Bloom's higher-level thinking skills – applying, analyzing, evaluating, creating. Fill in each box with a "1" – none or very little; "2" – moderate extent; "3" – significant extent.

	Applying	Analyzing	Evaluating	Creating
Apply knowledge of math, science, and engineering				
Design and conduct experiments				
Design a system, component, or process				
Function on multidisciplinary teams				
Identify, formulate, and solve engineering problems				
Understand professional/ethical responsibility				
Communicate effectively				
Understand global and societal context				
Recognize need for lifelong learning				
Knowledge of contemporary issues				
Use techniques, skills, and modern engineering tools				

3. Go online at www.engr.ncsu.edu/learningstyles/ilsweb.html and complete the "Index of Learning Styles Questionnaire" developed by Barbara Soloman and Richard Felder at North Carolina State University. Write a two-page paper on why it is beneficial for you to understand your preferred learning styles. Include changes in your behaviors you plan to make based on this new information.

4. Review the list of characteristics of expert learners presented in the Reflection exercise in Section 3.3. Divide the list into two categories: 1) those that describe you and 2) those that don't describe you. Take the list of characteristics that don't describe you and rank the items in importance. Pick the most important item and develop a written plan that includes steps you will take to move this item to the "describe you" list.

5. Consider the 80/20 rule described in Section 3.5. How does it fit with your past and current experiences? In high school, how many hours did you spend in class? How many hours did you spend learning outside of class? How does this compare to the 80/20 rule? Currently, how many hours do you spend in class? How many hours would the 20/80 rule suggest you should spend learning outside of class?

6. How does the statement that "learning is a reinforcement process" match your past learning experiences? Can you think of a specific example where you learned something through repeated reinforcement? Can you think of a specific example of something you would have liked to learn but didn't because of inadequate reinforcement? Write down five things you can do differently to take advantage of this idea. Implement what you wrote down.

7. Based on the five teaching styles presented in Section 3.4, determine how the way you prefer to learn compares to the way you are being taught. For each category, check the box that best describes the teaching style most prevalent in your math/science/engineering classes. Then check the box that describes the way you prefer to learn.

Type of Information Emphasized

Teaching style	Most prevalent in your classes	What you prefer
Concrete		
Abstract		

Mode of Presentation Stressed

Teaching style	Most prevalent in your classes	What you prefer
Visual		
Verbal		

Way Presentations Are Organized

Teaching style	Most prevalent in your classes	What you prefer
Deductive		
Inductive		

Mode of Student Participation

Teaching style	Most prevalent in your classes	What you prefer
Active		
Passive		

Type of Perspective Provided

Teaching style	Most prevalent in your classes	What you prefer
Sequential		
Global		

In what categories does the way you prefer to be taught *match* the way you are being taught? What are the implications for your learning? In what categories does the way you prefer to be taught *differ* from the way you are being taught?   What are the implications for your learning?

8. Using your favorite Internet search engine, research one of the teaching approaches mentioned in Section 3.5:

   - Interactive lectures
   - Problem-based learning
   - Inquiry-guided learning
   - Just-in-time teaching

   Write a one-page paper describing and critiquing the method.

9. Go to the Khan Academy website (www.khanacademy.org) and review the list of videos available there.   Pick two videos that address areas you could improve in.   Watch the videos and write a one-page statement about what you learned.

10. Go to the "Study Guides and Strategies" site (www.studygs.net).   Pick a topic that you would like to learn more about (e.g., active listening, influencing teachers, managing stress).   Study the material presented on that topic and prepare a 2-3 minute presentation on what you learned.   Find a willing student and deliver the presentation to him or her.

11. Go to the "University Survival" website (www.universitysurvival.com/student-success-skills).   Pick one of the following topics, read it, and prepare a 2-3 minute presentation for your *Introduction to Engineering* class on what you learned:

    - Managing your time
    - Overcoming your personal sound barrier to change your habits
    - Taking notes – math, science, and engineering classes
    - Knowing how to get things done
    - Understanding what constitutes academic dishonesty
    - Succeeding in an interview
    - Knowing how to build a network
    - Handling criticism

    Make a commitment to put what you learned into practice.

12. Complete the *Academic Success Skills Survey* at the end of this chapter.   Assign a point value to each question, based on the following point scale:

Strongly agree	+2
Agree	+1
Neutral	0
Disagree	-1
Strongly disagree	-2

    Compute your average score for the 16 statements in the survey.   Then rate yourself as "outstanding," "good," "fair," or "poor" in practicing good academic success skills.

13. Pick six of the 16 areas in the *Academic Success Skills Survey* you think are the most important for academic success.   What is your average score for these?

14. From the six academic success skills you identified as most important in Problem 13, pick the two skills you feel you most need to improve.   Develop a plan for improving in each area.   Implement the plan.

ACADEMIC SUCCESS SKILLS SURVEY

1. I interact regularly with my professors in positive, beneficial ways, both in and out of the classroom.

☐   ☐   ☐   ☐   ☐
STRONGLY   AGREE   NEUTRAL   DISAGREE   STRONGLY
AGREE   DISAGREE

2. I make effective use of my peers by regularly engaging in group study and collaborative learning.

☐   ☐   ☐   ☐   ☐
STRONGLY   AGREE   NEUTRAL   DISAGREE   STRONGLY
AGREE   DISAGREE

3. I schedule my time, utilizing time and priority management principles.

☐   ☐   ☐   ☐   ☐
STRONGLY   AGREE   NEUTRAL   DISAGREE   STRONGLY
AGREE   DISAGREE

4. I would give myself an A+ on the amount of time and energy I devote to my studies.

☐   ☐   ☐   ☐   ☐
STRONGLY   AGREE   NEUTRAL   DISAGREE   STRONGLY
AGREE   DISAGREE

5. I prepare for each lecture by reviewing my notes, reading ahead in the text, attempting some problems, and writing down questions.

☐   ☐   ☐   ☐   ☐
STRONGLY   AGREE   NEUTRAL   DISAGREE   STRONGLY
AGREE   DISAGREE

6. I keep up in my classes by mastering the material presented in the last class meeting before the next class meeting.

☐   ☐   ☐   ☐   ☐
STRONGLY   AGREE   NEUTRAL   DISAGREE   STRONGLY
AGREE   DISAGREE

7. I am aware of the importance of being immersed in the academic environment of the institution and spend as much time on campus as possible.

☐   ☐   ☐   ☐   ☐
STRONGLY   AGREE   NEUTRAL   DISAGREE   STRONGLY
AGREE   DISAGREE

8. I practice good study skills in areas such as note-taking and preparing for and taking tests.

☐   ☐   ☐   ☐   ☐
STRONGLY   AGREE   NEUTRAL   DISAGREE   STRONGLY
AGREE   DISAGREE

9. I am aware of the best methodologies for reading for comprehension and practice those methodologies during my learning process.

☐ STRONGLY AGREE   ☐ AGREE   ☐ NEUTRAL   ☐ DISAGREE   ☐ STRONGLY DISAGREE

10. I recognize the importance of goal setting and I have clear academic goals.

☐ STRONGLY AGREE   ☐ AGREE   ☐ NEUTRAL   ☐ DISAGREE   ☐ STRONGLY DISAGREE

11. I am effectively managing the various aspects of my personal life, such as interactions with family and friends, personal finances, and outside workload.

☐ STRONGLY AGREE   ☐ AGREE   ☐ NEUTRAL   ☐ DISAGREE   ☐ STRONGLY DISAGREE

12. I am highly motivated through a clear understanding of the rewards graduating in my chosen major will bring to my life.

☐ STRONGLY AGREE   ☐ AGREE   ☐ NEUTRAL   ☐ DISAGREE   ☐ STRONGLY DISAGREE

13. At my university, I know other students in my classes and feel part of an academic learning community.

☐ STRONGLY AGREE   ☐ AGREE   ☐ NEUTRAL   ☐ DISAGREE   ☐ STRONGLY DISAGREE

14. I am aware of and make optimal use of campus resources such as the writing center, counseling center, student health center, library, and career center.

☐ STRONGLY AGREE   ☐ AGREE   ☐ NEUTRAL   ☐ DISAGREE   ☐ STRONGLY DISAGREE

15. I feel good about myself and about my situation, and I am confident about my ability to succeed academically.

☐ STRONGLY AGREE   ☐ AGREE   ☐ NEUTRAL   ☐ DISAGREE   ☐ STRONGLY DISAGREE

16. I feel good about my institution and about the educational experience I am receiving.

☐ STRONGLY AGREE   ☐ AGREE   ☐ NEUTRAL   ☐ DISAGREE   ☐ STRONGLY DISAGREE

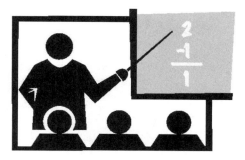

CHAPTER 4
Making the Most Out of How You Are Taught

You can observe a lot just by watching.

— Yogi Berra

INTRODUCTION

In Chapter 3, we provided an overview of how your teaching is delivered. In this chapter we will focus on how to take full advantage of that teaching process. By doing so, you will ensure you have a sound foundation on which to build your learning process. Specific ways to design your learning process are presented in the next chapter.

We begin this chapter by discussing early course preparation. We will emphasize that the *start* of a course is very important and that you need to be in the right courses, with the best available teachers, and have your textbooks and other materials. We will also discuss the course syllabus as a potential source of important information.

We then present a number of strategies and skills for taking full advantage of your lectures. One of the most powerful of these strategies is to prepare for lectures so that the lecture becomes a reinforcement of the material rather than an initial exposure to it. Other skills for getting the most out of your lectures are covered as well: skills such as listening, asking questions, and taking notes.

Next, we discuss strategies for making effective use of your professors, another important resource both in and out of the classroom. Too often students either overlook or fail to understand the many benefits their professors can provide them. After describing these benefits, we will teach you how to build the kind of positive relationships with your professors that you will need in order to obtain these benefits.

We close the chapter with a section on "Utilizing Tutors and Other Academic Resources." Taking advantage of these resources will require initiative on your part, but the benefits are well worth the effort.

4.1 EARLY COURSE PREPARATION

The beginning of a course can be likened to the beginning of a race. When the starting gun is fired, you have to be off and running. Otherwise you will spend the whole race trying to catch up – something you are unlikely to be able to do. You should be ready to "fire on all cylinders" from the get go. This means that you are in the right class, are mentally prepared, and have your textbook and other appropriate materials. Ideally, you will have reviewed your course selection with your academic advisor. This all begins with the process of selecting your courses, ideally through an academic advisement session with your advisor.

Being in the right class means that you have the necessary background and prerequisites and that the demands of the course, including its meeting time, fit into an overall manageable workload. When multiple sections of the same course are available to you, the selection of a specific section can be based on an evaluation of the various instructors. Sources of information about instructors include other students, other professors, and published student opinion survey results.

Don't Be Lulled into Complacency

Often a course starts slowly, with only a small amount of material presented in the first few classes. Don't use this as an excuse for getting off to a slow start. Remember how we likened the start of a course with the start of the race. Instead of being lulled into complacency, use the slow start to get on top of the course material. Later we will discuss the importance of mastering the material presented in each class before the next class comes. Make a resolve to adopt this approach from day one!

Mental preparation is not unlike getting psyched up for an important competition. As the start of the course approaches, check your mental frame of mind. Are you excited and focused? Are you clear that taking this class is important to you and you want to be there? If not, remind yourself of what you learned in Chapter 2 about why you want to be an engineer and review how this course fits into your roadmap for accomplishing that goal.

ACQUIRING TEXTBOOKS AND OTHER MATERIALS

You should get your textbooks right away – perhaps immediately after you register for a course. Don't wait until the term starts. You can benefit from scanning your textbooks and even studying the first few chapters during the break period preceding the start of the next term. Also, it is not uncommon for a campus bookstore to run out of books – and you can't afford to be without a book for the several weeks it might take additional texts to arrive. If you do buy your books early, save your receipts and refrain from writing in the books so you can return them if necessary. If money is a problem, consider buying used books either from your bookstore or from Internet book dealers such as Amazon.com or eBay.com (although used books may be more difficult to return).

Another choice to consider is ebooks. An increasing number of college textbooks are available in this format, although the more specialized the subject the less likely it will be available as an ebook. Among the advantages of ebooks are lower cost, live links, searchable text, and readability (e.g., font size can be adjusted). Among the cons are lack of availability and the need for a reader (Kindle, iPad, etc.). Perhaps the biggest issue is your personal preference. Do you prefer to read from a printed book or a digital display?

Make sure you have other materials you need as well. These would include a notebook for taking notes and other supplies and equipment such as a personal computer and/or a hand-held calculator.

USING THE COURSE SYLLABUS

Generally each of your professors will give you a course syllabus during the first week of the term. The syllabus can be a gold mine of valuable information. My advice is to study the syllabus thoroughly and keep it in a readily accessible place so you can revisit it frequently.

Sample Syllabus Content

Course Information
Course title, course number, credit hours, prerequisites, classroom location, dates and times class meets

Instructor Information
Full name, title, office location, office phone number, email address, office hours

Textbook(s)
Title, author, date (and edition), publisher, cost, extent to be used, other reference materials

Course Description/Objectives
Course description, instructional methods, content, goals, and objectives. Note: This item could range from as little as a repeat of the course description from the college catalog to a listing of detailed educational objectives, i.e., what students are expected to be able to do to demonstrate knowledge, skills, and attitudes learned in the course.

Course Calendar/Schedule
Daily (or weekly) schedule of topics, reading assignments, problem assignments, exam dates, due dates for assignments, special events (e.g., field trips, guest speakers, etc)

Course Policies
Attendance, lateness, class participation, missed exams or assignments, lab safety/health, emergency evacuation, academic dishonesty

Basis of Grading
Percentage of grade devoted to quizzes, final exam, homework, projects, essays and term papers, attendance, class participation

Available Support Services
Library references, learning center, computer resources

The syllabus should contain most or all of the above information. Since all of the items shown are things you need to know, if any are missing from the syllabus, I would encourage you to find them out by asking your professor.

Hopefully this list has persuaded you of the importance of mining the syllabus for important information.

4.2 PREPARING FOR LECTURES

Preparing for lectures is a powerful and effective strategy for success and an excellent opportunity to reinforce your learning. The idea of preparation is second nature for both engineering professors and engineering professionals. An engineering professor wouldn't think of coming to a lecture unprepared and a practicing engineer wouldn't think of coming to a meeting unprepared.

It is unfortunate that so few students prepare for lectures – or even know how to do it – for it yields so many benefits. It's a little like warming up for a physical workout. Students who take time to prepare for their lectures go into the lecture with more interest, follow the lesson with more ease, and come away with more knowledge than those who walk in cold.

All these benefits derive from preparation's role in the *reinforcement* process of learning. If you study a lecture topic in advance, even briefly, the lecture becomes your first reinforcement, rather than your initial exposure to the subject. Thus, both your level of learning <u>and</u> interest are enhanced.

In their excellent book *How to Study in College* [1], Walter Pauk and Ross J. Q. Owens put it this way:

> Each lecture can be viewed as a piece of a puzzle. The more pieces you have in place, the more you know about the shape and size of the pieces that remain. The course syllabus, the notes from your last lecture, and related reading assignments can all function as these puzzle pieces as you prepare for a lecture.

While preparing adequately for your lectures does require effort on your part, it's not all that difficult or time-consuming. Prior to class – the night before or, if feasible, during the hour just before class begins – review your notes from the previous class, read over the next section in your text, try a few of the problems at the end of the chapter, and write down questions about things you're unsure of.

Try to do this for at least a week or two to see how such little effort can have a big impact on what you get out of your lectures. I'm sure you'll be surprised by the results and subsequently make this part of your regular study routine.

REFLECTION

Think about going to a concert given by your favorite musical group. Which songs do you enjoy the most, those that you have heard many, many times before, or those you have never heard? Why do you think a person might enjoy or get more out of hearing songs they've heard before? Do you see how this relates to the idea of preparing for your lectures?

4.3 DURING YOUR LECTURES

Once you have prepared for a lecture, there are several tactics that will help you get the most out of the lecture: sit near the front, concentrate on the material being presented, practice good listening skills, take thorough notes, and ask questions.

SIT NEAR THE FRONT

Studies show that students who sit near the front of the classroom perform better than those who sit in the back. Sitting near the front has several obvious but important advantages. You will hear better, see better, have fewer distractions, and be better positioned if you want to ask a question or otherwise get your professor's attention.

"BE HERE NOW"

Getting the most out of your lectures requires that you learn how to keep your attention focused – i.e., that you "be here now." This is not easy, as most students – indeed, most people – have short attention spans. From time to time, your mind will wander to other thoughts. The result? You will tune out the lecture and perhaps miss important points.

When your mind wanders, you need to immediately "slap yourself" mentally and return your attention to the lecture. Every time you do this, you will increasingly strengthen your ability to concentrate on the "here and now." (You'll find this ability extremely valuable not only in lectures but in many other situations, both as a student and later as a practicing engineer. Just one example is the need to "stay on task" when working in study groups, an important success strategy we will discuss in Chapter 5.)

One last point. By all means turn off your cell phone and your laptop computer (unless you are using it to take notes). Having one or both of these devices on will make it even harder for you to "be here now."

LISTENING SKILLS

Good listening skills can be developed, but working to develop them is often overlooked. To a great extent, being a good listener means being an active listener. We tend to equate hearing with listening. Yet listening is much more than mere hearing. It is a conscious choice process. Once you know the difference between good listening habits and poor listening habits, you can choose one or the other. It's really up to you. Try them!

The following table contrasts nine characteristic of good listeners with those of poor listeners (adapted from *How to Study in College* [1]):

Poor Listener	Good Listener
Tunes out uninteresting and boring topics. Turns off quickly.	Works at finding value in all topics. Listens to discover new knowledge.
Tunes out if delivery is poor.	Judges value of the content rather than the delivery.
Listens for facts and details.	Listens for central themes. Uses them as anchor points for the entire lecture.
Brings little energy to the listening process.	Works hard at listening; remains alert.
Readily reacts with opposing views to new ideas. Starts listening to *themselves* when they hear something they don't agree with.	Focuses on understanding completely rather than coming up with opposing views.

Bothered by distractions.	Fights distractions; ignores bad habits of other students; knows how to concentrate.
Resists difficult material; prefers light recreational material.	Welcomes difficult material as exercise for the mind.
Interrupted by emotionally-charged words or ideas.	Does not get hung up on emotionally-charged words or ideas; listens with an open mind.
Daydreams and lets mind wander off with slow speakers or gaps in presentation.	Uses extra time to think more deeply about what the lecturer is saying; summarizes what has been covered.

REFLECTION

For each of the nine items in the table above, decide which column best describes you as a listener during your lecture classes. For each item in which you describe yourself as a "poor listener," decide whether you would benefit from changing your habit to one of a "good listener." Make a commitment to the change and try it out for a week in your classes.

NOTE-TAKING

Another effective way to get the most out of your lectures is to take good notes. Your notes essentially create a record of what your professor feels is important, and that in itself is important for two reasons. First, many professors cover only certain portions of a textbook while, second, others present material that the text does not address. In either case, your notes will help you know what to study for tests.

Tips or instructions on how to take good notes are difficult to give, for there is no one "correct" way to go about it. Your note-taking techniques will depend on a variety of factors, such as your own preferred style, the type of class, and the professor's teaching methods. The following generalizations might be helpful to keep in mind:

(1) Note only important details: Do not try to record everything the professor says.

(2) Include anything the professor writes on the board or conveys through visual aids (such as slides or overheads), for that usually signals "important details."

(3) Write down whatever you think you might encounter on the exam.

You might wonder whether to take notes in the traditional way (pen and paper) or take them on a laptop computer. Each has its advantages and disadvantages.

TAKING NOTES ON A COMPUTER. If you use a laptop or tablet computer, you can take notes using a word processor (such as Microsoft Word or Apple Pages) or note-taking software (such as Microsoft OneNote or Evernote).

Following are some **advantages** of taking notes on a computer:

- You will have a perfectly legible set of notes.
- You can keep all your notes in one place – organized and safe (provided you back them up).
- Sharing your notes electronically with others is easy.
- Your notes will be searchable.
- You can download information from websites or other electronic sources directly into your note file.
- You can move information around easily.

Following are some **disadvantages** of taking notes on a computer:

- You have to bring your laptop or tablet to class.
- You may type more slowly than you write. (Note: To improve your typing speed, try one of the many online programs such as www.keybr.com. It's free. And it's fun!)
- You have to worry about battery life and remember to back up your notes.
- You can only see a portion of your notes at one time.
- Note taking on the computer is not generally appropriate in courses that have lots of equations (i.e., most math/science/engineering courses).

After all these pros and cons, you may not have a choice if your professor does not permit laptops in class. The main reason for this would be to prevent students from playing computer games, surfing the web, checking email and Facebook, and the like during class. Just ask your instructor whether you can bring a laptop to class. If the answer is "yes," by all means refrain from doing anything other than using it to take notes.

TAKING HANDWRITTEN NOTES. Taking notes with pen and paper has its own considerations. One is whether to use a spiral notebook or a three-ring binder. Each offers advantages. With the three-ring binder, notes you take while reading the textbook, solutions to homework problems, and handouts and other reference materials can be easily integrated into your class notes. Another benefit is that you can spread your notes from a lecture out in front of you. On the other hand, if you use a spiral notebook, you're not likely to lose or misplace anything you have written in it.

LAYOUT OF NOTES. Whether you take notes manually or type them on a computer, it's important to lay out each page of your notes in a way that best facilitates your learning process. My suggestion is that you format each page to allow for three content areas:

1. Your actual note taking
2. Questions that your notes answer
3. A summary of the content of each page of notes

The Cornell Note-Taking Method, developed by Walter Pauk at Cornell University [1] and illustrated below, provides an excellent template with three structured areas – a place for your notes ("Note Taking Area"), a place for questions ("Cue Column"), and a place for a summary ("Summary Area"). A Microsoft Word template for Cornell Note Taking can be found at: www.timeatlas.com (Click on "5 Minute Tips").

If you prefer to take your notes manually, use a standard 8½" x 11" sheet of paper and divide it as the illustration below shows. Allow a 6" x 9" note-taking area, a 2½" margin on the left side for questions, and a 2" margin at the bottom for your summary.

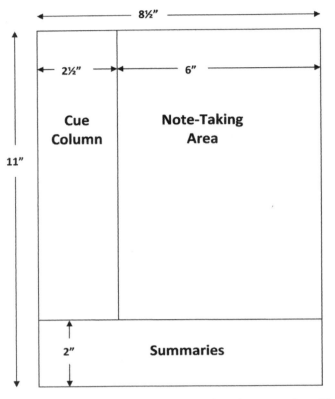

Take your notes as you normally would, but restrict them to the 6" x 9" note-taking area. Leave the cue column and the summary area blank. In Chapter 5 we will discuss how to use the blank areas as part of your learning process.

Whether you use the format above or just the old, reliable "fill-up-one-page-and-then-go-to-the-next-page" approach, remember that if you don't write something important down, it is unlikely you will be able to recall it later. Research in brain cognition has repeatedly shown that human memory is mostly short-lived. Unless an idea or information is consistently reinforced over a long period of time, it is quickly forgotten – usually in a matter of days. Your only alternative, then, is to record important information. That's why note-taking is an essential academic success strategy.

<u>**ONE LAST THOUGHT.**</u> More and more professors are turning to PowerPoint presentations for their lectures as an alternative to writing on a whiteboard or chalkboard. Ideally, you will be given copies of the slides, allowing you to concentrate mainly on listening to the lecture and jotting down important points rather than taking comprehensive notes. If you are not provided with copies of the PowerPoint slides, it will be virtually impossible for you to both transcribe the information on the slides and take notes on what the professor says. So do your best to capture the *main* ideas.

ASKING QUESTIONS IN CLASS

Another way to get the most out of your lectures is to ask questions. Don't be reluctant to do this. Many students are. When asked why, they give reasons like:

"I was afraid I would look stupid."
"I didn't want to bring attention to myself."
"I was too confused to know what to ask."
"I didn't want to take time away from the instructor or other students."

Don't be one of these students. If you're confused during a lecture, it's likely that many other students are as well. Just acknowledge it by raising your hand and, when called on, saying something like: "I'm confused about the last point you made." Take the view that the only dumb question is one that was never asked. And whose time do you think it is anyway? How about the view that it's your time? You deserve to ask questions!

While asking questions is a useful way to gain information in the classroom, improving your ability to ask good questions can benefit you in your career and all aspects of your life.

The importance of this subject was well-stated by the Greek philosopher Socrates more than 2,500 years ago:

> *The highest form of Human Excellence is to question oneself and others.*

Because of the importance of questioning skills, let's delve into the topic a bit more.

There are many ways to categorize questions. For our purpose, we can divide them into four types:

➢ Memory-level questions
➢ Convergent thinking questions
➢ Divergent thinking questions
➢ Evaluation thinking questions

Each of these types is related to a different way we receive and process knowledge.

Memory-level questions exhibit memory of previously learned material by recalling facts, terms, basic concepts, and answers. These questions usually begin with "Who," "What," "Where," or "When." Memory-level questions often solicit a "Yes" or "No" or very short answer. Examples might be:

- Who invented the electric light bulb?
- What chapters will be covered on the midterm?
- Where do I go to sign up for academic advising?
- When is the final exam?

Convergent thinking questions represent the analysis and integration of given or remembered information. These questions usually begin with "Why," "How," or "In what ways," and their answers involve explaining, stating relationships, and comparing and contrasting. Examples are:

- Why do we use AC current rather than DC current in our homes?
- How does a microwave oven work?
- In what ways does the Linux operating system differ from the Windows operating system?

Divergent thinking questions are those that require you to formulate answers by combining elements in a new pattern or proposing alternative solutions. These questions usually begin with "Imagine," "Suppose," "Predict," "If . . . then," "How might," "What are some of the possible consequences," "What could be changed to improve." These are often described as *open-ended* questions. Answers to these questions involve predicting, hypothesizing, inferring, reconstructing, or designing. Some examples:

- Suppose Einstein had never discovered his "Theory of Relativity." How would things be different now?
- How might you go about improving your writing skills?
- What could be changed to improve the gas mileage of SUVs?
- What are some of the possible consequences of global warming on engineering job opportunities in the future?

Evaluation thinking questions are those dealing with matters of judgment, value, and choice. These questions usually begin with "Defend," "Judge," "Justify," "What do you think about," "What is your opinion about." Examples might include:

- Could you justify why two years of calculus is required in the engineering curriculum?
- What do you think about the new drop/add policy?
- What is your opinion about the value of mandatory academic advising?

I hope these classifications will help you improve your skill at formulating and asking questions.

4.4 MAKING EFFECTIVE USE OF YOUR PROFESSORS

As discussed in the previous section, most of your professors are committed to a lecture style of teaching in which they convey knowledge to you in a one-way communication style. Most assign homework problems for you to do, collect and grade the problems, and so provide you with valuable feedback. Professors also determine your grade in the course – generally based on your scores on one or more tests and a final examination. This process of the professor lecturing, evaluating homework assignments, giving exams, and determining final course grades (which reflect students' mastery of the subject) is the standard teaching/learning process in engineering education.

IMPORTANT ROLES FOR YOUR PROFESSORS

But your professors can contribute much more than this to your overall education. The following is just a partial list of what professors can do for you:

- Give you the **benefit of the doubt** on a borderline grade.

- Provide you with invaluable **one-on-one instruction.**

- Give you **academic advising, career guidance,** and **personal advice.**

- **Monitor your progress** and **hold you accountable** for your performance.

- **Help you find a summer job** in industry and even **hire you** on their research grants.

- **Serve as a valuable reference** when you apply for jobs, either while you are a student or after you graduate.

- **Nominate you** for scholarships or academic awards.

REFLECTION

Reflect on the above list of important roles for your professors. Are these things you would like your professors to do for you? Which ones would be particularly important to you? Would you like to have a close advisor or mentor? Would you like to have one-on-one instruction from an expert? Would you like to have a future reference for a job or scholarship? What would it take on your part to ensure that your professors do these things for you?

VALUE OF ONE-ON-ONE INSTRUCTION. Of these roles, there is one in particular I'd like to expand on briefly, and that is one-on-one instruction – for this is probably the most valuable and beneficial role your professor can play outside of class.

One-on-one instruction is one of the best ways to learn, especially if the interaction is between an expert (i.e., teacher/professor) and novice (i.e., student). It is often referred to as the *Socratic method,* named after the Greek philosopher Socrates, who used this method when he taught over 2,500 years ago. The primary advantage of the Socratic method is that the teacher can know immediately if the student understands the subject of their dialogue and, if necessary,

adapt the lesson on the spot to ensure that the student truly learns it.

This teaching method would be ideal for engineering education – i.e., daily extended one-on-one meetings between just you and your professors – but realistically it is not possible. The most we can do is try to keep the teacher/student ratio as low as possible, while providing as many opportunities as possible for one-on-one instruction outside of the classroom or lecture hall.

One of these opportunities, and perhaps the best, is the weekly office hours that every professor is required to keep. In fact, the primary purpose of office hours is to give students the chance to work one-on-one with their instructors. If a student's schedule conflicts with his or her professor's office hours, as is often the case, most professors are willing to make appointments to meet with students at other times.

I urge you to use this opportunity regularly and frequently. As your education progresses, look for other opportunities to work one-on-one with your professors, such as offering to help them in their research projects or help out in their labs. Not only will such interactions enable you to learn more about engineering, you will establish the kind of relationships with your professors you need to derive the many benefits they can offer you.

TAKE RESPONSIBILITY FOR WINNING OVER YOUR PROFESSORS

To make effective use of your professors, you first must overcome any fear or intimidation of them you may feel. Being awed by your professors is a natural inclination since they are older and better educated, and they often project a confident "know it all" attitude. As a result, you may think that your professors don't care about you – or even that they are somehow "against" you. But this isn't true. After all, most professors chose an academic career because they like teaching and enjoy working with students.

Remember, too, professors are human beings just like you, and they have similar needs, fears, and insecurities as you do. They may very much need to be liked, want you to think they are good teachers, need to impress you with their knowledge, or fear that they might make a mistake and reveal that they don't have a total command of their subject matter.

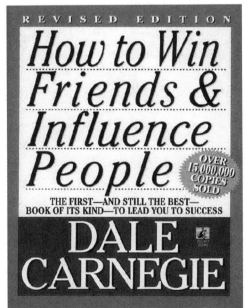

Once you get past any feelings of fear or awe, you need to realize that winning over your professors is <u>your</u> responsibility. <u>You</u> must take the initiative in establishing positive relationships with them.

HOW TO WIN OVER YOUR PROFESSORS. The real question is how can you go about winning over your professors so that they want to help you. Perhaps the "bible" for winning people over is the classic book by Dale Carnegie, *How to Win Friends and Influence People* [2]. Written in 1936, this book has stood the test of time and is still a best seller. I recommend it to you as an excellent resource to improve your "people skills."

Dale Carnegie's "Six Ways to Make People Like You" lists helpful strategies that you can use to win over your professors:

Six Ways to Make People Like You

Rule 1	Become genuinely interested in other people.
Rule 2	Smile.
Rule 3	Remember that a person's name is to him or her the sweetest and most important sound in any language.
Rule 4	Be a good listener. Encourage others to talk about themselves.
Rule 5	Talk in terms of the other person's interest.
Rule 6	Make the other person feel important – and do it sincerely.

Dale Carnegie's book is filled with anecdotes. Most are dated, but their messages are timeless. The one I like the best is this story:

> C. M. Knaphle, Jr., of Philadelphia, had tried for years to sell coal to a large chain-store organization. But the chain-store company continued to purchase its fuel from an out-of-town dealer and continued to haul it right past the door of Knaphle's office. Mr. Knaphle made a speech one night before one of my classes, pouring out his hot wrath upon chain stores, branding them a curse to the nation.
>
> And still he wondered why he couldn't sell to them.
>
> I suggested that he try different tactics. To put it briefly this is what happened. We staged a debate between members of the course on "Resolved that the spread of the chain store is doing the country more harm than good."
>
> Knaphle, at my suggestion, took the negative side; he agreed to defend the chain stores, and then went straight to an executive of the chain-store organization that he despised and said, "I am not here to try to sell coal. I have come to you for help because I can't think of anyone else who would be more capable of giving me the facts I want. I am anxious to win this debate, and I'll deeply appreciate whatever help you can give me."
>
> Here is the rest of the story in Mr. Knaphle's own words:
>
> *I had asked this man for precisely one minute of his time. It was with that understanding that he consented to see me. After I had stated my case, he motioned me to a chair and talked to me for exactly one hour and forty-seven minutes. He called in another executive who had written a book on chain stores. He wrote to the National Chain Store Association and secured for me a copy of a debate on the subject. He feels that the chain store is rendering a real service to humanity. He is proud of what he is doing for hundreds of communities. His eyes fairly glowed as he talked, and I must confess that he opened my eyes to things I had never even dreamed of. He changed my whole mental attitude. As I was leaving, he walked with me to the door, put his arm around my shoulder, wished me well in my debate, and asked me to stop in and see him again and let him know how I made out. The last words he said to me were: "Please see me again later in the spring. I should like to place an order with you for coal."*
>
> *To me that was almost a miracle. Here he was offering to buy coal without my even suggesting it. I had made more headway in two hours by becoming genuinely interested in him and his problems than I could have made in ten years by trying to get him interested in me and my coal.*

I'm sure you get the point of this story; more importantly, I hope it has given you ideas on how to approach your professors. The anecdote and Dale Carnegie's "Six Ways to Make People Like You" emphasize the importance of both showing interest in others and approaching them from their perspective.

REFLECTION

How do you relate to Mr. Knaphle's story? Have you had experiences where you came at people from your side of things and didn't get what you wanted from them? Have you ever tried approaching someone from their side of things? Who was it? Teacher? Parent or close relative? Friend? Co-worker? Boss at work? How could you apply the lessons of Mr. Knaphle's story to interacting effectively with your professors?

CHARACTERISTICS OF YOUR PROFESSORS YOU CAN COUNT ON. Just as Dale Carnegie knew Mr. Knaphle could win over the chain store executive by appealing to his interest in promoting chain stores, there are three characteristics of professors that you can almost always count on and therefore use to win them over:

(1) Professors think their areas of technical specialty are critically important and extremely interesting (and they have stories they love to tell about how they got involved in their specialty).

(2) Professors have elected an academic career over professional practice and they believe they are outstanding teachers.

(3) Professors aren't called "professors" for nothing. They have big intellects and lots of knowledge, and they love to convey what they know to others.

Your challenge as a student is to avoid doing anything that conflicts with these characteristics of professors; rather, think of ways to interact with your professors that tap into these characteristics.

BEHAVIORS TO AVOID. We could make a long list of behaviors that conflict with professors' belief in the importance and interest of their technical specialties:

- Coming late to class
- Yawning or sleeping in class
- Talking in class
- Doing homework in class
- Using a cell phone in class to search the Internet, play video games, or text
- Leaving class early
- Failing to do the assigned homework

I'm sure you can add to this list.

The above behaviors also conflict with professors' belief that they are good teachers, as do other behaviors such as:

- Correcting professors' mistakes in an antagonistic tone
- Complaining that exams are too hard
- Complaining that grading is unfair
- Sending non-verbal messages to your professors that you dislike them personally

REFLECTION

Reflect on the list of "behaviors to avoid." Do you engage in any of these behaviors? If so, make a commitment to stop them. Are there additional behaviors to avoid that should be on this list? Make a commitment to stop those as well.

WINNING BEHAVIORS. Given what you now know, a good way to win over your professors is to send them messages that you find their subject both interesting and important, and that you value them as a teacher. You can start by practicing the opposite of the behaviors listed above. Be on time to class. Sit in the front. Pay attention. Ask questions. Volunteer to answer questions the professor asks the class. Apply yourself to the assigned homework.

There is also a much more direct way. Just tell them! In my experience, professors get far too few compliments. I'm not sure why students are so reluctant to tell their professors that they like the course, are interested in the subject, or appreciate the good job the professor is doing in teaching the class. I can assure you that doing so will go a long way toward winning over your professors.

One additional strategy for developing a positive relationship with your professors: show interest in them. In my *Introduction to Engineering* class, I assign students to visit one or more of their professors during their office hours and ask them questions like: "Where did you go to college?" "How did you choose your technical specialty?" "How did you decide to become a teacher?" Students report very positive experiences from such interactions. Try it!

And a final bit of advice. Make the effort to know your professors' names. One of Dale Carnegies "Six Ways to Make People Like You" is:

> ***Remember that a person's name is to him or her the sweetest and most important sound in any language.***

I have frequently encountered students who couldn't tell me the name of their professors. Please don't be such a student.

MAKE SURE YOUR PROFESSORS KNOW YOUR NAME.

I'd like to raise one more issue related to names. Do your professors know your name? Probably not. Professors are busy people generally with too many students to learn and retain all of their names. But they do know some of their students' names and why shouldn't you be one of those? Set a goal of having your professors know your name and take responsibility for making it happen.

REFLECTION

What benefits will come to you if your professors know you by name that might not come to you if they don't? Do your professors in your key classes know your name? If you're not sure, how could you find out? If they don't know your name, what steps can you take to ensure that they do?

UNDERSTANDING WHAT YOUR PROFESSORS DO

You might benefit from understanding what university professors do. By doing so, you can be more sensitive to the demands placed on them and be more effective in building relationships with them.

University professors do much more than teach classes. In fact, they are expected to perform in three primary categories:

- Teaching
- Research
- Service

The **teaching category** includes not only classroom teaching but also course and curriculum development, laboratory development, academic advising, and supervision of student projects or theses.

The **research category** includes creating and organizing new knowledge; disseminating and organizing new knowledge through publication of research papers, textbooks, software, and presentations at scholarly meetings; participation in professional societies and other activities that keep the faculty member up-to-date technically; and generating funds to support research.

The **service category** may include community involvement, participation in faculty governance through service on university committees, public service, consulting, and a variety of other activities.

Although most universities expect a faculty member to demonstrate accomplishments in all three categories, the relative importance given to each varies from one university to the next, depending on the characteristics of the institution. At one end are the so-called "research universities," which emphasize success in creating new knowledge, publishing the results, and obtaining funds to support these activities. Teaching loads at these institutions are relatively light – usually one or two courses a term.

At the other end of the spectrum are predominately undergraduate universities, which emphasize teaching. At these institutions, the teaching load is generally heavier – usually three to four courses a term. Some research or equivalent professional activity also is expected of faculty members, but much less than at research institutions.

COMMUNICATING WITH PROFESSORS BY EMAIL AND TEXT MESSAGING

Until recently, most communication between students and their professors occurred either in class (just before class, during class, or just after class) or during face-to-face meetings in the professor's office. Nowadays, email (and to a lesser extent texting) has become a powerful tool for communication between students and their professors.

Emailing Your Professors. Because of its potential for misuse and abuse, it is particularly important that you make judicious use of email. If your professor puts his or her email address in the course syllabus, you can assume they will be receptive to your messages. If not, you should ask the professor whether they welcome e-communications from students.

An article in the *New York Times* about emailing professors [3] starts out:

"One student skipped class and then sent the professor an email message asking for copies of her teaching notes. Another did not like her grade, and wrote a petulant message to the professor. Another explained that she was late for a Monday class because she was recovering from drinking too much at a wild weekend party."

Hopefully, you wouldn't be inclined to send similar messages to your professors. Here are some guidelines that might help you with the protocols for emailing your professors.

- **Write from your college or university email account** – Shows that your email is legitimate and not spam.

- **Include the course number in your subject line** – Relieves your professor of the chore of figuring out what class you're in.

- **Use an appropriate greeting** – "Hi/Hello/Dear Professor _____" is always appropriate. Don't use "Hey _____." Avoid first names unless you have been specifically invited to do so. Avoid "Dr." unless you are sure the professor has a doctoral degree.

- **When you get a reply** - Express your gratitude. Just hit "Reply" and write "Thank you" or a little bit more if appropriate. Try to avoid sending a second message that requires the professor to respond again.

- **Things to avoid**

  Rote apologies (better to relate any sad or serious circumstances in person)

  Unexpected attachments

  Criticism of the professor or other students

  Requests for information you can find from other sources

  Email abbreviations and jargon you might use with a friend

  Emoticons

  Exclamation points

  Quotes from famous people in your signature line

  Anything you would not say in a face-to-face meeting with your professor

  Making unreasonable demands on your professor's time

- **Things to do**

  Be clear, concise, and polite

  If you are emailing with a problem, suggest a solution

  Capitalize and punctuate correctly

  Proofread what you've written

  Sign with your full name, course number, and meeting time

Although wide-scale adoption of email started over 20 years ago, it is still developing as a tool for communication between students and their professors, and both groups are still learning how to best use email to mutually benefit the teaching/learning process. You can make a positive contribution to this process by following the guidelines above.

TEXTING YOUR PROFESSORS. Although texting is a common means of communication among friends and family, it is not widely used for communication between students and professors. It is unlikely that your professors will provide you with their cell phone numbers and encourage you to text them. But you certainly can ask whether they are open to it. Don't be surprised, however, if the answer is "no." If your professor does invite you to text him or her, use this privilege sparingly, perhaps only when time is of the essence. Receiving numerous text messages can be onerous and distracting. And unless your professor has purchased a texting plan, he or she will incur a charge for each text message you send.

4.5 UTILIZING TUTORS AND OTHER ACADEMIC SERVICES

Your university or college offers a number of academic services to support your education. Examples of these are tutoring, recitation sections of large lectures, and other academic services. These services are generally free to you, since you pay for them in advance in your tuition and student fees. However, receiving the benefits of these academic services almost always requires that you take the initiative. They will not seek you out. Part of good academic gamesmanship, then, is for you to find out about the resources available to you and make optimal use of them. Remember, you paid for them!

TUTORING

Tutors are an excellent source of the type of one-on-one instruction discussed previously in this chapter. Some students are reluctant to utilize tutors, equating the need for tutoring with an admission they are doing poorly or need help.

After what we've said about the myth of "succeeding on your own," you should realize how unfounded and counterproductive such reluctance is. If, however, you find yourself in this bind – in need of help but resistant to seek tutoring – try looking at tutoring in a more positive light: as an opportunity for you to have a dialogue with an expert on a subject you want to learn.

Your university may provide tutoring services through a variety of sources. Tutoring may be available through a campus-wide learning assistance center. Your mathematics department may run a math lab. Members of your engineering honor society, Tau Beta Pi, may provide tutoring as a service to the engineering college.

If free tutoring is not available, you might find listings for tutors available for hire at your career center. You could also just ask an upper-class student to help you. Lots of students like to "show off" their knowledge.

RECITATIONS/PROBLEM-SOLVING SESSIONS

A common teaching mode is large lectures (100 to 300 students or more) accompanied by small recitations. Recitations are generally taught by graduate students serving as teaching assistants. The purpose of the recitation is to amplify and reinforce the main concepts from the lecture and to work on problems. While there is a limited opportunity for you to ask questions in large lectures, you will find more than ample opportunity to ask questions in recitations. To get the most out of recitations, it is important that you have studied the material presented in the lecture and attempted the assigned problems.

OTHER IMPORTANT ACADEMIC RESOURCES

Among the other academic resources available to you, here are some you should certainly look into and use.

- The **academic resource center** not only provides tutoring in math and science; it can also help you improve your reading, writing, and study skills.

- Your **university library** is certainly a source of important books, periodicals, online material, and other references to support your engineering education. It also holds workshops and seminars on how to access all these sources of information. Reference librarians are available through email, telephone, or in person to assist students individually in identifying and retrieving information for research, study, or personal use.

- **Student computer labs** provide access to computer hardware and applications software, access to the Internet, resource materials, and training.

- **Academic advising** gives you help in reviewing your overall academic plan and your progress in following it as well as guidance in selecting courses for the next term. More information on this important academic resource can be found in Chapter 8.

- The **university/college catalog** is your "bible" of important academic information such as rules and regulations, college and department information, curricular requirements, and course descriptions.

- Your **registrar's office** can help you with various academic procedures, including changing majors, dropping and adding classes, enacting/ processing grade changes, and transferring course credits from other institutions.

To find out what your campus offers and where these services can be found, check your university or college webpage, catalog, or department or school student handbook. Freshman orientation programs are also helpful for learning about student services. And don't forget one of your best resources: other students.

SUMMARY

In this chapter, we presented strategies and skills for making the teaching process work for you. We began by likening the start of a course to the start of a race, pointing out that you need to be ready to go as soon as the "starting gun is fired" and discussing ways to ensure this happens.

Then we discussed a number of important strategies and skills for taking full advantage of your lectures. One of the most powerful of these is to prepare for lectures so that the lecture becomes a reinforcement of the material rather than an initial exposure. Other skills for getting the most out of your lectures, such as good listening, effective note-taking, and asking good questions in class, were covered as well.

We next explored the contributions that faculty can make to the quality of your education, both in and out of the classroom. We explained that deriving these benefits is <u>your</u> responsibility to pursue and presented a variety of strategies and approaches for you to take in order to establish the kind of positive relationships with your professors you will need to receive these important "extras."

In addition to the support that your professors can offer you, we listed other academic resources that can provide you equally valuable support. But, once again, <u>you</u> must assume responsibility for seeking them out and taking advantage of them.

In the next chapter, we focus on how to make the learning process work for you. If you have implemented the skills and strategies presented in this chapter, you should have a strong foundation on which to build that process.

REFERENCES

1. Pauk, Walter and Owens, Ross J.Q., *How to Study in College, 10th Edition*, Wadsworth Publishing, New York, NY, 2010.

2. Carnegie, Dale, *How to Win Friends and Influence People*, pp. 65-66, Simon and Schuster, New York, NY, 1936.

3. Glater, Jonathan D., "Emailing Professors: Why It's All About Me," *New York Times*, February 21, 2006.

PROBLEMS

1. In the section on early course preparation, an analogy was made between the start of a course and the start of a race. Think about the idea that the start of things can be very important. Make a list of situations in which the start is extremely important. For each item on the list, consider whether a poor start can be overcome and what it would take to do so.

2. How do you decide which section to take in courses in which multiple sections are offered? Do you work proactively to get in the section with the best teacher? What do you do? What are some strategies you could adopt that you haven't been using?

3. Review the course syllabus in at least one of your key courses. Compare the information there to the list of items presented in Section 4.1 on "Using the Course Syllabus." Are most or all of the items included? If not, which ones are missing? Take on the task of filling in the missing items.

4. Conduct an Internet search on the subject "ebooks versus print books." Read several of the articles you find and make a list of the pros and cons of ebooks. Make another list of the pros and cons of print books. Do you prefer using ebooks or print books? Has this exercise influenced your view toward either? Are your current textbooks available as ebooks?

5. Make a commitment to prepare for your lectures using the approach discussed in Section 4.2 for a two-week period. At the end of the two weeks, write a two-page paper discussing the benefits you received (or didn't receive) from doing this.

6. Review the characteristics of good and poor listeners presented in Section 4.3 on "Listening Skills." Pick two or three items from the "poor listener" list that describe you, if only occasionally. Make a commitment to try to change the habit to that described in the "good listener" column. Try out the new habit for two weeks and reflect on how it worked.

7. Weak typing skills can limit your productivity during both college and your career. On a scale of zero to ten, how would you rate your typing skills? Go to <u>www.keybr.com</u> and take

the typing test there several times to get a baseline on your typing speed and accuracy. If your speed is less than 45 words per minute, make a commitment to practice on this website until your speed reaches this level.

8. Try out the Cornell Note-Taking System presented in Section 4.3. Check if your bookstore sells pre-made forms for this purpose. If not, make up your own forms. Preview the information on studying and annotating your notes in Section 5.2 of the next chapter. Make a commitment to follow the process of annotating your notes, including developing questions in the "cue column" and summarizing each page in the "summary area." Try answering the questions in the cue column out loud over a period of two weeks and reflect on how well it worked.

9. Practice your questioning skills by making up three additional questions that would be appropriate to include as problems at the end of this chapter. Email them to your *Introduction to Engineering* course instructor and to me at: rlandis@calstatela.edu.

10. For one of your key classes, make up two questions that fit each of the four categories of questions presented in Section 4.3 (memory level, convergent thinking, divergent thinking, evaluation thinking). Take them with you to the next lecture and see if you have an opportunity to ask one or more of them.

11. Pick one of the important academic success skills below and conduct a search on that skill using your favorite Internet search engine.

    Note-taking skills
    Listening skills
    Questioning skills

    Gather information from a number of sites and write a 250-word paper on what you learned about the skill.

12. Make a list of behaviors that would send signals to your professors that you don't think their technical specialty is either interesting or important. Do you engage in any of these behaviors? Which ones?

13. Explain how the skills you develop in learning how to make effective use of your professors will directly carry over in the engineering work world.

14. Do you think that grading is objective or subjective? Ask two of your professors how they go about making up their final grades. Ask them what factors they consider in deciding borderline grades. Are these factors objective or subjective? Write a one-page "opinion report" on what you learned.

15. Go see one of your professors during his or her office hours. Ask one or more of the following questions:

    a. Why did you choose teaching as a career rather than professional practice? Would you recommend an academic career to others? Why or why not?

    b. Would you advise me to continue my engineering education past the B.S. degree? What are the advantages of getting an M.S. degree? A Ph.D. degree?

    c. I understand that your technical specialty is in the field of _____. How did you get interested in that field? Do you think it would be a good field for me to consider?

    d. What do you think are the most important factors in an engineering student's academic success?

16. Make up five additional questions like the ones in Problem 15 that you could ask one of your professors. Pick the three you like the best and ask them of one of your other professors.

17. Study the "protocols for emailing your professors" in Section 4.4. Do you agree with these guidelines? When you email your professor, do you follow them? Write down reasons why each of the "things to avoid" are on the list.

18. Look into the availability of free tutoring services on your campus. Are there tutors to help you with your mathematics classes? Are there tutors to help you write a term paper?

19. Pick one of the following offices on your campus. Stop by and seek information about the academic services offered there. (Note: The specific names may vary from campus to campus.)

Academic Resource Center
Library Reference Desk
Open Access Computer Laboratory
Registrar's Office
Academic Advising Center

Prepare a two-minute presentation on what you learned for your next *Introduction to Engineering* class meeting.

Making the Learning Process Work For You

To improve is to change. To be perfect is to change often.

— Winston Churchill

INTRODUCTION

In Chapter 3, we provided overviews of how your teaching is delivered and how your learning takes place. In Chapter 4, we presented approaches for taking full advantage of the teaching process. This chapter focuses on designing your learning process.

We begin the chapter by discussing two important skills for learning – reading for comprehension and analytical problem solving. Your effectiveness as a learner will depend to a great extent on how well you develop your skills in these two areas.

Next we examine powerful principles and approaches for organizing your learning process. The importance of keeping up in your classes is discussed and steps for mastering the material presented in each class are presented. Mastering the material presented in a class before the next class comes requires a strong commitment to both time management and priority management, so we explore these important topics in detail.

We then describe approaches for preparing for and taking tests. Since the primary way you will be called on to demonstrate your learning will be through your performance on tests, it is imperative that you excel in this area.

The chapter concludes with an in-depth discussion of making effective use of one of the most important resources available to you: your fellow students. Working collaboratively with your peers, particularly in informal study groups, sharing information with them, and developing habits of mutual support will be critical factors in your academic success and the quality of education you receive.

5.1 SKILLS FOR LEARNING

We begin this chapter by discussing two important skills for learning: (1) reading for comprehension and (2) analytical problem solving.

READING FOR COMPREHENSION

Much of your learning will depend on how well you understand information presented in written materials. And you can expect that much of this material will be highly technical in nature. In fact, a typical four-year engineering curriculum will include about three years (96 semester credit hours) of technical courses (math, physics, chemistry, engineering, computing) and one year (32 semester credit hours) of non-technical (humanities, social science, communication skills) courses.

Although the methodology you learn in this section can be applied to both your technical and non-technical courses, it is particularly important for your technical courses.

One important difference in reading technical material is "speed." You may equate having "good" reading skills with being a fast reader and even consider taking a "speed reading" course. Speed reading may be helpful in reading a novel for pleasure or for reading the morning newspaper, but trying to read too fast may work against you in your technical courses. Mastering mathematics, science, and engineering content is generally a slow, repetitive process that requires active participation on your part.

There are a number of methodologies for reading for comprehension. All involve developing your skills in three areas:

- What you do *before* you read

- What you do *while* you read

- What you do *after* you read

BEFORE YOU READ. The amount you learn from a reading task can be greatly enhanced by taking a few minutes to do three things before you start reading:

1. **Purpose** – Establish a purpose for your reading. The purpose might be entertainment or pleasure. Or it might be to find out one single piece of information. However, in technical courses, more often than not the purpose is to comprehend principles and concepts that will enable you to solve problems at the end of a section or chapter in a textbook.

2. **Survey** – Decide on the specific scope/size of the reading (one page, one section, one chapter). Devote a few minutes to survey/skim/preview that page, section, or chapter. Look at headings and subheadings. Inspect drawings, diagrams, charts, tables,

figures, and photographs. Read the introductory section or paragraph and the summary section or paragraph.

3. **Question** – Write down questions that you want to answer from the reading. A useful technique is to turn section headings and subheadings into questions. For example, some questions you might have written down for this section are: What do I need to learn to do before I read? What are the benefits of doing these things? What are some of the things that might keep me from doing them?

WHILE YOU READ. Following is a list of suggestions for improving your comprehension when reading technical material.

• Never sit down to study without a paper and pen or pencil at hand. You'll need them for sketching graphs, checking derivations, summarizing ideas, and raising questions. This approach to active reading is very important.

• Focus on concepts, not exercises or problems. The goal of most technical coursework is to enable you to understand concepts that can be applied to a variety of problems. Rather than focusing on how one particular problem is solved, first aim to understand the general concepts thoroughly. Pay close attention to the mathematical formulas. Work carefully through each derivation. Take time to absorb graphs and figures. One of the biggest mistakes students make is to skim over the material in order to get on with the homework problems. Don't truncate the learning process in your rush to get an assignment done.

• Don't try to read too quickly. In a half hour, you might read 20-60 pages in a novel. But expect to spend the same half hour on just a few lines of technical material. (Mathematics says a lot with a little!) Become an active participant in your learning process. At every stage, decide whether the concept presented was clear. Ask questions. Why is the concept true? Do I really understand it? Could I explain it to someone else? Do I have a better way to explain it?

• Write down anything that you don't understand. Where possible, frame it in the form of a question. Seek to answer such questions by re-reading the text or using alternate sources such as other textbooks or the Internet. Pose the questions to other students or to your instructor during his or her office hours. As you read, there may be questions that pique your curiosity but are not answered in the reading material. Take on the challenge of finding the answers to these questions.

• Periodically stop reading and **recite** what you have read. Using your own words, repeat to yourself (ideally aloud) what you have read. This is perhaps the most important step in the reading process. If the material is difficult, you may want to recite after reading just one paragraph. By reciting, you implant the knowledge in your brain. Do not look at the book as you recite. If you can't remember what you read, reread the material.

AFTER YOU READ. Once you have read your text, most of the learning is still ahead of you. The following are three important tasks to perform after your reading: recite and reread, review, and solve problems.

- **Recite and Reread**. – Recall that during the process of preparing for your reading, you formulated questions you would like to have answered by the reading. Recite answers to those questions. If you need to, reread sections of the text. Again, it is preferable to recite aloud. Even better, recite to others. One of the best ways to learn anything is to teach it to someone else. Form a study group or meet with a study partner and practice teaching each other what you have learned from the reading (more detail on group study is covered in Section 5.4). You can even tell friends and family members what you are learning. Talking about what you have learned is a powerful way to reinforce it.

- **Review** - Recall from Chapter 3 that learning is a reinforcement process. Only through repeated exposure to information can we move it from our short-term memory to our long-term memory. Plan to do your first complete review within one day. Review the important points in the text and recite some of the main points again. Do it again in a week and then when you prepare for a quiz or exam and again when you prepare for the final examination.

- **Solve Problems** - Once you have read your text for comprehension, it's time to work problems. Being able to solve problems with speed and accuracy is to a great extent what you will be judged on in your math/science/engineering coursework. This requires both a systematic problem-solving approach and lots and lots of practice. You can't work too many problems. First do any assigned problems. Don't stop there though. If possible, work all the problems in the book. If you still have time, work them again.

Most of the problems you will encounter in math/science/ engineering coursework can be described as "analysis" problems. Because of the importance of this type of problem in your education, the next section discusses a methodology for analytical problem-solving.

Before you read on, go to the next page and try the exercise to illustrate the principles of reading for comprehension.

PROBLEM-SOLVING

Engineers are problem-solvers. Much of your engineering education, and indeed your engineering career, will center on improving your ability to think both logically and creatively to solve problems. There are many types of problems. Some examples would be:

Type of Problem	Example
Mathematics problems	Determine the probability that two students in a class of 30 have the same birthday.
Science problems	Find a theory that explains why a dimpled golf ball travels further than a smooth one.
Engineering analysis problems	Find the stress in a beam, the temperature of an electronic component, or the voltage at a node in a circuit.

Engineering design problems	Design a better mousetrap, a car alarm that goes off if the driver falls asleep, a better device for waiters and waitresses to carry plates of food from the kitchen to the table.
Societal problems	Propose a systematic solution to global warming, immigration, health care, or world hunger.
Personal problems	Solve health problems, financial problems, or relationship problems.

EXERCISE

Apply the reading methods you have just learned to the following passage:

Shortcut For Adding Consecutive Numbers Starting with 1

The following theorem was published in Levi Ben Gershon's 13th century manuscript *Maaseh Hoshev (The Art of Calculation)*:

"When you add consecutive numbers starting with 1, and the number of numbers you add is odd, the result is equal to the product of the middle number among them times the last number."

In the language of modern-day mathematics, this is written as:

$$\sum_{i=1}^{2k+1} i = (k+1)(2k+1)$$

Before reading, did you establish a purpose? Did you first skim the passage? Did you write down any questions you wanted answered? During reading, did you follow any of the suggestions made in the previous section? Most importantly, did you recite what you learned? Can you apply the theorem? For example, how quickly can you add all consecutive numbers from 1 to 99? What level of understanding did you achieve? What does the symbol "k" represent? Can you find a better way to represent the theorem mathematically? Were any questions that piqued your curiosity? Do you want to know more about the life and work of Levi Ben Gershon? Did learning about this theorem interest you in other similar theorems?

GENERAL PROBLEM-SOLVING METHODOLOGY. Because there are so many types of problems, there are many different problem-solving methodologies. For example, you undoubtedly learned in high school about the scientific method in which a hypothesis is developed to explain some observed physical phenomenon and then tested through experiment. You also learned about the engineering design process in Chapter 2 of this book. Both are specific problem-solving methodologies.

A general approach for solving problems involves the following steps:

- Figure out where you are (problem definition).
- Figure out where you want to be (e.g., customer need or business opportunity).
- Determine what resources are available.
- Identify any constraints.
- Develop possible solutions that could solve the problem while staying within available resources and not violating any of the constraints.

- Choose the best solution.
- Implement it.

If you master this problem-solving approach, you will not only be able to use it in your work as an engineering professional, but also in all aspects of your personal life. It can be applied to such varied problems as buying a new TV, dealing with your car breaking down in the middle of nowhere, finding a job, running for president of your engineering student organization, starting your own company, or stopping the re-zoning of an area of your neighborhood.

All of these examples and most real engineering problems can be described as open-ended problems, meaning that they have no single right answer or solution.

ANALYSIS PROBLEMS. Much of your engineering education, however, particularly in the first several years, will not deal with open-ended problems but rather will focus on "analysis problems." Generally, analysis problems have one single right answer. They typically involve translating a physical problem into a mathematical model and solving the resulting equations for the answer. The problem statement will be provided to you by your instructor either in the form of a handout or a problem from your textbook. The principles you need to solve the problem will typically be contained in your text material, although you may need to draw on knowledge from prerequisite courses. Your success in engineering study will depend to a great extent on your ability to solve such problems accurately and often under time pressure.

Real-world problem-solving is, for the most part, not a science but rather an art. It involves learning, thinking, logic, creativity, strategies, flexibility, intuition, and trial and error. Even so, becoming a proficient analytical problem-solver can best be accomplished if you adopt, practice, and become proficient at the following four-step systematic approach (adapted from famous mathematician George Pólya) [1]. This approach is not an <u>algorithm</u> (i.e., a series of steps that if applied correctly are guaranteed to lead to a solution) but rather a *heuristic* (a general set of guidelines for approaching problem-solving that do not guarantee a solution).

Step 1: Understand the problem. Read the problem carefully. Identify the question you are being asked to answer. Identify the unknown(s) and assign each unknown a symbol. List all known information. Draw a figure, picture, or diagram that describes the problem and label it with the information you have extracted from the problem statement.

Step 2: Devise a plan. The goal of this step is to find a strategy that works. Because there are many possible approaches, this is perhaps the most difficult step in the problem-solving process. Think about possible relationships between the known information and the question you need to answer. Depending on the nature of the problem, the following is a list of problem-solving strategies to try:

Solve a simpler problem.
Make an orderly list.
Look for a pattern in the problem.
Draw a diagram.
Use a model.
Use a formula.
Work backwards.
Make a table.
Guess and check.
Eliminate possibilities.
Solve an equation.

Use direct reasoning.
Consider special cases.
Think of a similar problem.
Solve an equivalent problem.

Step 3: Carry out the plan. Implementing the plan depends on the nature of the problem and the problem-solving strategy chosen. In all cases, work carefully and check each step as you proceed. By the time you reach this step, you should have reduced the problem to a purely mathematical one. Work through each step of any mathematical manipulations or derivations. Complete any required calculations using an electronic calculator or computer. Take particular care to ensure correct handling of units, a frequent source of errors in engineering problem-solving.

Step 4. Look Back. Examine the solution you obtained. Make sure it is reasonable. Recheck your calculations and review your reasoning. Verify that your answer is consistent with the information given.

As you become better and better at solving problems, let's hope you don't reach the point shown in the following cartoon.

© Randy Glasbergen
www.glasbergen.com

"MEMO: It has come to my attention that every time we solve one problem, we create two more. From now on, all problem solving is forbidden."

5.2 ORGANIZING YOUR LEARNING PROCESS

In the following sections you will learn about ways to organize your learning process.

"TAKE IT AS IT COMES"

I use the expression "take it as it comes," an axiom you've undoubtedly heard before, to emphasize what I consider to be the **KEY** to success in mathematics, science, and engineering courses. Stated more explicitly:

> **Don't allow the next class session in a course to come without having mastered the material presented in the previous class session.**

Because, to me, this is the single most powerful academic success strategy, *if you are willing to put only one new behavior into practice, <u>this is the one to choose</u>!*

Have you ever wondered why a typical course is scheduled to meet only one, two, or three times a week rather than all five days? And why the total weekly hours of class meetings are so limited? After all, if you met nine hours a day for five straight days, you could theoretically complete an entire course in one week and cut down the time to complete your undergraduate degree from four years to less than one year!

The answer is obvious. The teaching part of the teaching/learning process could be compressed into one week but certainly not the learning part. You can only absorb a certain amount of material at one time, and only when that material is mastered can you go on to new material. Thus, your institution has designed a sound educational system in which professors sequentially cover small amounts of material for you to master. However, unless you do your part, you can easily turn that sound educational process into an unsound one.

REFLECTION

Reflect on your approach to your classes. Do you operate on the principle that you master the material presented in each class before the next class comes? Or do you put off studying until a test is announced and then cram for it? If you are in the second category, are you willing to change? What do you need to do to make that change?

PROCRASTINATION

Most students make the mistake of studying from test to test rather than from class to class. In doing so, they fall victim to a student's greatest enemy – procrastination.

Procrastination is an attitude that says, "Do it later!" "Doing it later" rarely works in any course, but especially not in math, science, and engineering courses, in which each new concept builds on the previous ones.

If you are a procrastinator, for whatever reason, you are likely to ignore the sequential nature of engineering study, somehow thinking that you have the capability to absorb complex information all at once. So you can't realistically expect to succeed if you delay your studying until a test is imminent. That's why I tell you to "take it as it comes."

A Common Trap

One trap you can fall into is a false sense of security because the teacher presents the material so clearly that you feel you understand it completely and therefore do not need to study it. But when you attend a lecture that is presented clearly, it only proves that the <u>teacher</u> understands the material.

What is necessary is for <u>you</u> to understand it – for you to be able to give the lecture. In fact, that should be your goal in every class – to get to the point where you could give the lecture.

Often I am invited to be a guest speaker at *Introduction to Engineering* courses that use this text. Because it's such an important topic, I always make it a point to address the subject of procrastination. Typically, I'll ask the class: "How many of you would say – 'I am a procrastinator'?" To this day, I am surprised that not only do almost all the hands go up, but they go up enthusiastically (almost proudly). It seems as though there's pride in being a member of an undesirable club – the "Procrastinator Club."

You may think of procrastination as doing nothing. But we're always doing something. Ultimately, ***procrastination involves choosing to put off something we know we should be doing and instead doing something we know we shouldn't be doing.*** And why would we do such a thing? Why would we delay an action we know we should be doing? There are lots of reasons:

Fear of failure – Task is perceived as too difficult: "If we don't attempt it, then we haven't failed."

Fear of success – Accomplishing task might be resented by others, or success might bring responsibilities and choices that we view as threats or burdens.

Low tolerance for unpleasant tasks – Task is viewed as not being enjoyable. Doing the task may bring some discomfort.

Disorganized – Prefer to spend time worrying about <u>not</u> doing rather than doing. Unwilling to set priorities, develop a schedule, and stick to the schedule.

There are many strategies for dealing with procrastination. If you find you're putting something off because you think of it as unpleasant, a good approach is the ten-minute rule. Acknowledge, "I don't feel like doing that," but make a deal with yourself and do it for ten minutes anyway. After being involved in the activity for ten minutes, then decide whether to continue. Once you're involved, it's easier to stay with a task.

And if you're overwhelmed with the difficulty of a task, use the "Swiss Cheese Method." Poke holes in the big project by finding short tasks to do that will contribute to completion of the larger project.

Procrastination is a big subject, with many complete books devoted to it. If you are a chronic procrastinator, you might want to learn more about the subject by reading one or more books about it, e.g., [2,3,4].

REFLECTION

Recall one or two recent examples where you put off a task you should have done. Describe each task you put off. What were you <u>thinking</u> when you chose to put off the task? What were you <u>feeling</u> when you were procrastinating? Can you identify any discomfort you were avoiding by putting off the task?

MASTERING THE MATERIAL

As previously discussed, you will learn better if you "take it as it comes," mastering the material presented in each class session before the next session comes. In fact, research on learning indicates that the sooner you study after your initial exposure to the material, the more fortified your learning will be. Having a study session right after class would be ideal, but if that's not possible, doing it the same day would be better than the next day. Since your goal is to <u>master</u> the material, start by studying and annotating your notes, reading (or rereading) the relevant portions in your text, and working problems – as many as you can.

<u>**LEARNING FROM YOUR LECTURE NOTES**</u>. Recall in the section on the **Cornell Note-Taking Method** (Chapter 4, Pages 117-118), it was recommended that you structure your notes to leave two areas blank for use during your learning process:

(1) Cue Column
(2) Summary Area

Now we will describe how you can take advantage of this new way of taking notes by adopting a systematic process for learning from them. The goal of this process is to increase your understanding of what was covered in class and to move as much as possible from your short-term memory to your long-term memory through repetition, review, and reinforcement.

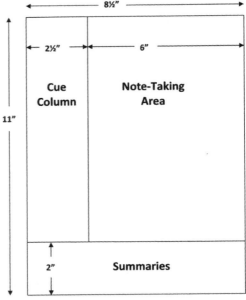

The process of learning from your notes involves six separate but interrelated steps.

<u>**Step 1 - Study and annotate your notes**</u>. Read/Go over each page of your notes and fill in any missing information. Add words, phrases, facts, or steps in a derivation you may have skipped or missed, and fix any difficult-to-decipher jottings. As you study your notes, enliven them by making liberal use of colored pens or pencils. Highlight important points by underlining, circling or boxing, or using arrows.

<u>**Step 2 – Question or Reduce**</u>. Formulate a question answered by each major term or point in

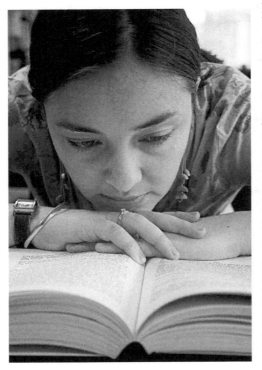

your notes and write it in the Cue Column. This is a little bit like the TV game show *Jeopardy* where the contestants are given the answer and asked to supply the question. An alternative approach would be to reduce each main idea or set of facts into a key word or phrase and write it in the Cue Column.

<u>**Step 3 – Summarize**</u>. Write a summary of each page in the Summary Area at the bottom. Summarizing forces you to think about the broader context of the lecture. Your summary should answer the questions: "What is this page about?" and "How does it fit into the day's lecture?" These summaries will be particularly helpful in finding key information when you are studying for an exam.

<u>**Step 4 – Recite**</u>. Once you have studied and annotated your notes and filled in the Cue Column and the Summary Area for each page, it is time for the most important step in your learning process: ***recitation.*** The process of reciting is relatively straightforward. Go back to the first page and cover the Note-Taking Area with a

blank sheet. Read the first question or key word in the Cue Column. If you wrote questions, answer each question in your own words. If you wrote key words, describe the main idea or set of facts referred to. Ideally, you should recite out loud. If you are reluctant or unable to do this, recite by writing out your answers. Slide the blank sheet down to check your answer. If your answer is wrong or incomplete, try again. Continue this process until you have gone all the way through your notes.

Step 5 – Reflect. After you have completed the first four steps above, take some time to reflect on what you have learned. Ask yourself questions like, "What's the significance of what I have learned?" "How do the main ideas of this lecture fit together into a bigger picture?" "How do they fit into what I already know?" "What are some possible applications of the key ideas from this lecture?" "Which ideas are clear?" "Which ideas are confusing?" "What new questions do the ideas raise?"

Step 6 – Review. Working through the process above will not only increase the amount you learn from your lectures and notes, it will also convert your lecture notes into study notes for future reviews. I suggest you review all your notes once each week. Doing so won't take much time but will pay off immensely in the long term. You'll find that if you spend just ten minutes to review your notes weekly, you'll retain most of what you initially learned. Then give your notes a more thorough review when you prepare for a test and then again when you prepare for the final exam. And don't forget to use reciting to reinforce what you have learned during each review session.

READING/REREADING THE TEXT. Next, read or re-read (*re-read* if you read the text to prepare for the lecture) the text material. Follow the "Reading for Comprehension" methodology described in Section 5.1. Resist your urge to skim over the material in order to get to assigned problems. Read or re-read to understand the concepts. And make reciting a key part of your reading process.

SOLVE PROBLEMS. As previously discussed, solving one or two problems, even if that's all your professor assigns, will not ensure an adequate level of understanding. If time permits, work <u>all</u> of the problems in the book. If more time is available, work them a second time. Practice, practice, practice! The more problems you solve, the more you will learn. Remember:

> *Much, perhaps most, of the learning in math, science, and engineering courses comes not from studying or reading but from solving problems.*

To the extent possible, utilize the analytical problem-solving methodology described in Section 5.1. By doing so, you'll improve your problem-solving capability over time.

After you feel you have mastered the material, you can reinforce your understanding through a group study session or by going to visit your instructor during office hours to address specific questions or problems.

Only then will you be ready for your next class meeting. At that point, you will have reinforced your understanding of the material several times. Later you will again reinforce it when you review for a test and still later when you prepare for the final exam.

LEARN TO MANAGE YOUR TIME

Time is an "equal opportunity" resource. All people – regardless of their socioeconomic status, gender, ethnicity, physical challenges, cultural practices, or any other kind of "difference" – have exactly the same amount of time. Everyone, including you, gets 168 hours each week – no more, no less.

> *There is no point in saying that you have no time,*
> *because, you have just as much as everyone else.*

Time is an unusual and puzzling concept. Even the most brilliant scientists and philosophers aren't sure how to explain it. But we do know some things about it. It can't be saved. When it's gone, it's gone. It also seems to pass at varying speeds – sometimes too slowly and other times too quickly. It can be put to good use, or it can be wasted. Some people accomplish a great deal with their time, while others accomplish virtually nothing with theirs.

People who accomplish a great deal, without exception, do two things:

(1) They place a high value on their time.

(2) They have a system for scheduling and managing their time and tasks.

Some of these systems are very sophisticated, and you may wish to look into acquiring one, particularly when you become a practicing engineer. As a student, you can do quite well with a monthly calendar to note your appointments and a simple form for making a day-to-day schedule. (See form on Page 160).

HOW MANY HOURS SHOULD YOU STUDY? Once
you commit to staying on top of your classes and reinforcing your learning as often as possible, you must make sure to allot a sufficient number of study hours to truly master the material covered in a one-hour lecture. Earlier, in presenting the "60-Hour Rule" (See Chapter 1, Page 29), we mentioned the standard rule-of-thumb that you should study two hours out of class for every hour in class. But this is often a gross oversimplification or, at best, a very limited generalization. In actuality, the amount of study time required will vary from course to course, depending on such factors as:

- How difficult the course is
- How good a student you are
- How well prepared you are for the course
- What grade you want to receive

For demanding technical courses, it is doubtful that two hours of studying for every hour spent in class are enough. The appropriate number for you may be three, four, or even five hours. Although this may be difficult to assess, especially early on in your education, it's good to determine a number for each of your classes. You can always adjust it later.

REFLECTION

On a scale of zero to ten, how good of a student are you? Reflect on each of your courses. How difficult is the course (very difficult, moderately difficult, not difficult)? How well prepared are you for the course? What grade do you want in the course? Based on this information for each of your courses, write down how many hours you need to study for each hour of class time.

Once you have decided that for a particular course you should study, say, three hours between one class meeting and the next, and you have blocked out a schedule for studying as soon after each lecture as possible, you have done the easy part. The hard part is actually doing it. Putting these approaches into practice requires you to be organized and skilled in managing your time.

MAKING UP YOUR WEEKLY SCHEDULE.

Your effectiveness and productivity as a student will be greatly enhanced by scheduling your time. The approach I took when I was a student was to sit down each Sunday night with a form like the one shown at the end of this chapter (Page 160) and schedule my entire week. You may find that a whole week is too much, and prefer to schedule a day or two at a time. That's fine. The idea is to find a scheduling method that works for you.

For whatever time period you choose, first write down all your commitments: classes, meetings, part-time work, time to get to and from school, time for meals, and so forth. The rest of your time is available for one of two purposes: study or recreation.

Next, schedule blocks of time to study. You have already decided how much study time you need between one class meeting and the next, and you know the advantages of scheduling this time as soon after each class meeting as possible. Write down both <u>where</u> and <u>what</u> you will study. Students tend to waste too much time between classes making three decisions:

(1) Should I study now or later?
(2) Where should I study?
(3) What should I study?

By making these decisions in advance, you will eliminate this unnecessary waste of time.

EXERCISE

Do you schedule your study time? If so, how is it working for you? If not, why don't you? Sit down with a form like the one on Page 160 of this chapter and schedule your time for the next week. After scheduling your commitments (class, work, appointments, etc.), schedule your study time following the principles presented in this section. Include information on both where you plan to study and what course you will focus on during each time block. Make a commitment to follow your schedule for the next week.

Once your study time is scheduled, check to see that you've left open time for breaks, recreation, or "down time." If not, you are probably overcommitted. You have taken on too much. One of the advantages of making a schedule is that it gives you a graphic picture of your situation. Remember, don't "program yourself for failure." Be realistic about what you can handle. If you are overcommitted, you should probably let something go. Try to reduce your work hours, your extracurricular activities, or the number of units you are taking.

MAKE A SERIOUS COMMITMENT TO YOUR SCHEDULED STUDY TIME. Making up a weekly schedule, you'll find, is easy and fun. But sticking to it will be a challenge. The key is to make a serious commitment to your study time. I'm sure you take your class time as a serious commitment. If, for example, five minutes before a class a friend asked you to go have a cup of coffee or a Coke, you would say, "Sorry. I can't because I have a class." But what about your study time? What if the same friend came up to you just as you were about to go to the library to study?

You need to make the same commitment to your scheduled study time as you do to your class time. After all, much more learning occurs out of class than in. It always astonishes me that students are so willing to negotiate away their study time. Every time you put off an hour of studying, you are giving up time that you cannot recapture, and that means borrowing time from the future. If, however, your future is already scheduled, as it should be, the notion of borrowing time from the future is impossible. You're talking about time that won't be there.

To monitor yourself, outline the hours you actually study in red on your schedule form. At the end of each week, you will be able to readily count up how much studying you did. If you are doing poorly in your classes, I'll bet you will see a direct correlation between your performance and the amount of studying you are doing.

Initially, you may find that you have made a schedule you are unable to follow. Don't beat yourself up over that. But, more importantly, don't use it as an excuse to give up scheduling your time completely. Over time, you will learn about what you can and cannot do and become more proficient at scheduling your time.

Predictable Outcome of Scheduling Your Time

If you are like most students, you will find that by scheduling your time and following the schedule, you will feel as though you have more time than you did before and your stress level will go down. Many students spend more time worrying about the fact that they are not studying than they do actually studying. "Tending to business" can give you a real sense of well-being.

In summary, the benefits of scheduling your study time are:

- You will be able to see immediately if you are overextended.

- You are more likely to keep up in your classes and to devote adequate time studying.

- You'll get immediate feedback as to how much you are actually studying.

- You'll learn about yourself – both what you can and cannot do.
- You'll feel that you have more time than you ever had before.
- You'll feel much less stressed out over school.

DAILY PLANNING: "TO DO" LIST.

One final approach to getting the most out of each day is to make up a daily "to do" list. To do this, take a few minutes each evening and write down a specific list of what you want to get done in the next day.

Then prioritize the items on the list, ranking them from top to bottom or classifying each as high, medium, or low priority. The next day, work on the most important items first. Try to avoid the urge to work on items that are easy or fun, but are of low priority. As you complete items, cross them off your "to do" list. At the end of the day, evaluate your progress and reschedule any items that remain on your list. Once again, though, if you repeatedly find that you can't accomplish everything on the list, you are probably over-scheduling yourself. And having to reschedule unaccomplished "to do" items means borrowing from the future, time that isn't there.

USING A LONG-TERM PLANNER.

In addition to planning each week, you need a way to keep track of long-term commitments, important dates, and deadlines. Your campus bookstore or a local office supply store has both academic year planners and calendar year planners for this purpose.

Enter appointments, activities, events, tasks, and other commitments that extend beyond the current week in this planner. These might be academically-related, such as test dates, due dates for laboratory reports or term papers, meetings of student organizations, engineering seminars or guest speakers, and advising appointments. Also include personal appointments such as medical and dental checkups and car maintenance schedules; special occasions such as birthdays, anniversaries, and holidays; and recreational activities such as parties, concerts, plays, and the like.

Each week, as you make up your weekly schedule, transfer commitments from your long-term planner to your weekly schedule.

You may consider keeping your weekly schedules and long-term planners so that in the years to come you can enjoy them as a reminder of what you did during this uniquely important period of your life.

PRIORITY MANAGEMENT.

If you want to move to a higher level of managing your life, Stephen Covey points the way in his powerful book *Seven Habits of Highly Effective People* [5]. Covey's guiding principle is to:

> ## *Organize and execute around priorities.*

Priority management means doing what needs to be done. There are two dimensions to deciding what needs to be done:

- How urgent is it? (Requires immediate attention or doesn't require immediate attention)
- How important is it based on your personal values? (Important; or not important)

These two dimensions – urgency and importance – are frequently confused. It's almost second nature to think that anything that's urgent must be important. The phone rings, we have to answer it. Our favorite TV show is on, we have to watch it. A friend wants to talk, we have to talk. Urgent matters press on us. They *demand* our attention. They're often popular to others. And often they are pleasant, easy, fun to do. But many urgent matters are not important!

"Importance" relates to whether it needs to be done at all. Not important should mean we don't do it at all. Much of our time and effort is devoted to tasks that are not important, whether they are urgent or not.

These dimensions can be shown visually by the following four quadrant matrix:

I **Urgent** **Important**	II **Not Urgent** **Important**
III **Urgent** **Not Important**	IV **Not Urgent** **Not Important**

Key to the process of priority management is the criteria we use to determine what is important. This depends on our value system. Suffice it to say, candidates for high personal value include:

- School
- Family
- Friends
- Health
- Personal goals

Stay out of Quadrants III and IV. People who spend time almost exclusively in Quadrants III and IV lead basically irresponsible lives. Effective people stay out of Quadrants III and IV because, urgent or not, activities in these quadrants aren't important. You might have already guessed that staying out of Quadrant III will require you to become good at saying "no."

Activities in Quadrant I are what Covey describes as "crisis management" activities. Much of your life is dominated by activities that are both urgent and important. Time to go to class. Need to prepare for tomorrow's exam. Term paper due. Need to go to work. All important and urgent things.

We can't ignore the urgent and important activities of Quadrant I. However, our overall effectiveness will be controlled by Quadrant II – i.e., how we handle the things that are important but don't have to be done today. Since we can't skip Quadrant I activities, finding time for Quadrant II activities will require that we give up activities from Quadrants III and IV.

One bit of good news: In time, choosing Quadrant II activities will have the benefit of reducing the need to always operate from the crisis management perspective of Quadrant I.

REFLECTION

Reflect on the academic success strategies presented in the chapters indicated:

Structuring your life situation (Chapter 1)
Preparing for lectures (Chapter 4)
Seeking one-on-one instruction from your professors (Chapter 4)
Utilizing tutors and other academic resources (Chapter 4)
Scheduling your study time (Chapter 5)
Mastering the material presented in each class before the next class comes (Chapter 5)

Are these Quadrant I, II, III, or IV activities? Do they have to be done immediately or not? Are they important or not?

5.3 PREPARING FOR AND TAKING TESTS

As you learned in Chapter 1, a vital component of successful engineering study is becoming a *master* at preparing for and taking tests.

PREPARING FOR TESTS

Clearly, the best way to prepare for tests is to practice the many strategies discussed earlier. When I hear a student boast that he or she stayed up all night studying for a test, I know this is a student who is not doing well. This is substantiated by a recent study by researchers at the UCLA Semel Institute for Neuroscience and Human Behaviors [6] that concluded:

"Sacrificing sleep for extra study time is counterproductive."

You, too, should recognize this by now.

The student who brags about staying up all night to study most likely does not study from class to class, does not schedule his or her time well, does not understand the learning process (i.e., the need for incremental, reinforced learning), and does not realize the pitfalls of studying alone (The image of a student staying up all night studying for a test certainly fits the "lone wolf" metaphor, doesn't it?).

The truth is, if you have incorporated the study skills we have discussed into your regular study habits – even just the one skill of "taking it as it comes" – preparing for a test is not very hard. It merely involves adjusting your schedule several days prior to the test to review the material. You should never have to cover new material when preparing for a test.

There is, however, one major aspect of test-taking that distinguishes it from all other forms of studying and learning: time pressure. That is, to do your best on tests, you need to learn how to work under the pressure of time.

Here are some useful tips that will improve your performance on tests and lessen your anxiety about taking them. Several days before a test, spend a portion of your study time working problems under a time limit. If you can, obtain tests from previous semesters or, better yet, construct your own. Creating and taking your own practice exams will give you invaluable experience in solving problems under pressure, plus it will give you the added advantage of learning to "scope out" tests. In time you will significantly improve your ability to work under pressure and to predict what will be on tests.

Unlike the student who stays up all night frantically cramming, be sure to get eight or nine hours of sleep before a test. Arrive at the test site early so you have ample time to gather your thoughts and be sure you have whatever materials you'll need: paper, pencils, allowed reference material, and acceptable computation tools. A certain amount of psyching yourself up, similar to what an athlete does prior to a big game, might be helpful; however, you don't want to get so nervous that you can't concentrate.

TEST-TAKING STRATEGIES

When you are given the test, don't start work immediately. Glance over the entire test first and quickly separate out the easier problems from the harder ones. Many instructors grade on a curve, which means that your grade will be based on its relation to the class's average performance, not your individual score alone.

If this is the case, you also need to size up the overall difficulty of the test and guess what the class average will be. In fact, jot down your estimate so that you can compare it later with the actual outcome. Through this process, over time you will become adept at sizing up tests. You will be able to recognize that on one test, it may take a score of 90 to get an "*A*," while on another test it may only require 50. Knowing that you only need to get a portion of the problems correct for a good grade will greatly affect the way you approach a test.

Once you have sized up the test, don't start with the first problem; start with the easiest one. As you work the easier problems and accumulate points, your confidence will build and you will develop a certain momentum.

But always keep an eye on the clock. If you divide the time available by the number of problems, you will know approximately how much time to spend on each.

Use this as a guide to pace yourself. Also, try to complete a problem before leaving it, and avoid jumping from one uncompleted problem to another, since you will waste time getting restarted on each.

Although you are under a time constraint, be sure to work carefully and attentively, as careless mistakes can be very costly. It is probably smarter to work three of five problems carefully than to do all five carelessly.

And by all means, never leave a test early. What do you have to do that could be more important than achieving the highest possible score on a test? If you have extra time, check and recheck your work. No matter how many times you proofread a term paper, mistakes can still be overlooked. The same is true for a test.

5.4 MAKING EFFECTIVE USE OF YOUR PEERS

We close this chapter with one of the most important academic success strategies: making effective use of your peers. Your peers can significantly influence your academic performance, either positively or negatively.

Negative peer pressure put on those who apply themselves to learning is an age-old problem. Derisive terms like *dork, wimp, nerd, geek,* and *bookworm* are but a few of those used to exert social pressure on the serious student. You may have experienced this type of peer pressure in high school if your friends were not so serious about their academics as you, and you may have been forced into a pattern of studying alone – separating your academic life from your social life.

> *As we discussed in Chapter 3, the "lone-wolf" approach to your academics may have worked for you in high school, but it is doubtful that it will work for you in engineering study, where the concepts are much more complex and the pace much faster.*

Even if you are able to make it through engineering study on your own, you will miss out on many of the benefits of collaborative learning and group study.

OVERVIEW OF COLLABORATIVE LEARNING

In a previous section, we discussed teaching modes: large lectures, small lectures, recitations, and tutoring sessions. Now we turn to *learning modes*. There are really only two:

(1) Solitary

(2) Collaborative

Either you try to learn by yourself or you do it with others.

As I travel the country, I always make an effort to visit *Introduction to Engineering* classes, where I ask students, "How many of you, when you study, spend some of that time with at least one other student?" Generally, in a class of 30 students, three or four hands will go up. Then I ask, "How many of you spend all of your time studying by yourself?" And the remaining 90 percent of hands go up.

> *My anecdotal research indicates that about 90 percent of first-year engineering students do virtually 100 percent of their studying alone.*

Hence, the predominant learning mode in engineering involves a student working alone to master what are often difficult, complex concepts and principles and then apply them to solve equally difficult, complex problems.

The fact that most students study alone is indeed unfortunate because research shows that students who engage in collaborative learning and group study perform better academically, persist longer, improve their communication skills, feel better about their educational experience, and have enhanced self-esteem. We just read essentially the same message in that excerpt from the Harvard University study (See Chapter 3, Page 103). As even more evidence, Karl A. Smith, Civil Engineering professor at the University of Minnesota and a nationally recognized expert on cooperative learning, has found that [7]:

> **Cooperation among students typically results in:**
> a. **Higher achievement and greater productivity**
> b. **More caring, supportive, and committed relationships**
> c. **Greater psychological health, social competence, and self-esteem**

Over a period of many years, I have made a special effort to understand why most first-year engineering students study alone. Whenever I have the opportunity, I ask students, "Why don't you study with other students?" I almost always get one of these three answers:

(1) "I learn more studying by myself."

(2) "I don't have anyone to study with."

(3) "It's not right. You're supposed to do your own work."

The first of these reasons is simply wrong. It contradicts all the research that has been done on student success and learning. The second reason is really an excuse. Your classes are overflowing with other students who are working on the same homework assignments and preparing for the same tests as you are. The third reason is either a carryover from a former era when the culture of engineering education emphasized "competition" over "collaboration," or it comes from that romanticized ideal of the "rugged individualist" that we debunked in Chapter 3.

Today, the corporate buzzwords are "collaboration" and "teamwork" and engineering programs are under a strong mandate to turn out graduates who have the skills to work well in teams.

If you are using any of these reasons to justify your "lone-wolf" approach to academic work, you should now see their inherent problems and consider changing your approach.

BENEFITS OF GROUP STUDY

If you're still not convinced, then look at the issue from a different perspective. Instead of focusing on the weaknesses or problems of solitary study, consider the strengths or benefits of group study. In this light, you will find three very powerful, persuasive reasons for choosing the collaborative approach over the solitary one:

(1) You'll be better prepared for the engineering work world.

(2) You'll learn more.

(3) You'll enjoy it more.

Each of these is discussed in the following sections.

YOU'LL BE BETTER PREPARED FOR THE ENGINEERING WORK WORLD. Whether you choose to study alone or with others often depends on what you think is the purpose of an engineering education. If you think the purpose of that education is to develop your proficiency at sitting alone mastering knowledge and applying that knowledge to solving problems, then that's what you should do. However, I doubt you will find anyone who will hire you to do that. It's not what practicing engineers do by and large.

So if you spend your four or five years of engineering study sitting alone mastering knowledge and applying that knowledge to the solution of problems (and perhaps becoming very good at it), you will have missed out on much of what a quality engineering education should entail.

A quality education trains you not only to learn and to apply what you learn, but also to communicate what you know to others, explain your ideas to others, listen to others explain their ideas to you, and engage in dialogues and discussions on problem formulations and solutions. You may come up with a very important "breakthrough" idea, but if you can't convince others of it, it is unlikely that your idea will be adopted.

YOU'LL LEARN MORE. Do you recall our earlier discussion of traditional teaching modes, all of which keep learning to a minimum? In essence, group study and collaborative learning pick up where those modes leave off – and the result is an <u>increase</u> in what you learn.

There are a number of ways to explain how this happens. One is the adage that *"two minds are better than one."* Through collaborative study, not only will more information be brought to bear, but you will have the opportunity to see others' thought processes at work. Perhaps you have played the game *Trivial Pursuit*. It always amazes me how a small group of people working together can come up with the answer to a question that no member of the group working alone could have done.

Another explanation comes from the claim that:

> *If you really want to learn a subject, teach it.*

It's true! As an undergraduate engineering student, I took three courses in thermodynamics. Yet I didn't really understand the subject until I first taught it. When two students work together collaboratively, in effect, half the time one student is teaching the other and half the time the roles are reversed.

YOU'LL ENJOY IT MORE. Group study is more fun and more stimulating than solitary study, and because you'll enjoy it more, you are likely to do more of it. This wonderful benefit of group study can be illustrated by the following personal story.

My Own Experience with Group Study

When I was working on my Ph.D., a close friend of mine and I took most of our courses together. To prepare for exams, we typically would meet early on a Saturday morning in an empty classroom and take turns at the board deriving results, discussing concepts, and working problems. Before we knew it, eight or ten hours would have passed.

There is no way I would have spent that amount of time studying alone on a Saturday at home. Would you? The temptations of TV, the Internet, phones, email, and friends, along with the need to run errands or do work around the house, would surely have prevailed over my planned study time. By integrating my academic work with my social needs, I enjoyed studying more and did more of it.

The true value of academic relationships is illustrated in the following cartoon:

Campus radical.

FREQUENTLY ASKED QUESTIONS ABOUT COLLABORATIVE LEARNING

Once students embrace the concept of collaborative learning, they generally have questions on how to make it work. The three most frequently asked (and probably most important) questions are:

- What percentage of my studying should be done in groups?
- What is the ideal size of a study group?
- What can be done to keep the group from getting off task?

Although there are no definitive answers to these questions, the following points serve as reliable guidelines.

PERCENTAGE OF TIME.   Certainly, you should not spend all of your study time working collaboratively.   I would suggest somewhere between 25 and 50 percent.   Prior to coming together, each member of a group should study the material and work as many problems as possible to gain a base level of proficiency.   The purpose of the group work should be to reinforce and deepen that base level of understanding.   The better prepared group members are when they come together, the more they can accomplish during their study sessions.

SIZE OF STUDY GROUP.   When you hear the term "study group," what size group do you think of? Five? Ten? Fifteen? My ideal size is *two*. Think study "partners." When two people work together, it is easier to maintain a balanced dialogue, where each is the "teacher" for half the time.

Triads can work well too.   In larger groups, however, it can be difficult to ensure equal participation and members often feel the need to compete for their fair share of the time. Even between study partners, a conscious effort may be required to keep one of the two from dominating the dialogue.   My advice is to keep the groups small.   If more people come together to study, it's okay.   Generally, subgroups of twos or threes will develop.

While I think two is the ideal size of a study group, Richard Felder of North Carolina State University sent me the following different view:

> "With two people you don't get sufficient diversity of ideas and approaches, and there's no built-in mechanism for conflict resolution, so the dominant member of the pair will win most of the debates, whether he/she is right or wrong. Five is too many – someone will usually get left out. I suggest three as the ideal size, with four in second place, and two in third."

I would encourage you to experiment to see what works best for you.

STAYING ON TASK.   You may find it difficult to stay on task when working with others. There are no simple solutions to this problem, for it really boils down to students' discipline and commitment to their education.   Once again, though, size may be a factor: The larger the group, the more difficult it will be to keep everyone focused on academics.   Yet even in groups of two or three, staying on task can be a problem.

I have found it helpful to split up a group's meeting time into a series of short study sessions with breaks between sessions. Agree, for example, to study for 50 minutes and then take a ten-minute break. After the break, it's back to work for another 50 minutes, followed by another ten-minute break. And so on.

If nothing else seems to help your group to stay on task, then you're left with only one solution: *Just do it.*

REFLECTION

Do you spend some fraction of your study time on a regular basis with at least one other student? Or do you spend virtually 100 percent of your study time alone? If you don't study with other students, why not? Did the ideas presented in this section persuade you of the value of group study and collaborative learning? Are you willing to try it out? Make a commitment to identify a study partner in one of your key classes and schedule a two-hour study session with that person.

It Really Works

I often conduct workshops on collaborative learning, and at some point I have half of the class work on a problem in small groups and the other half work by themselves on the same problem. After about ten minutes, the ones who are working alone start looking at their watches and appear restless and bored. When time is called after 45 minutes, those who are working in groups are disappointed and ask for more time. They often express that they are just getting "hot" on a solution to the problem.

The next day I ask, "How many of you continued thinking about, working on, or talking to others about the problem we did yesterday?" Most of those who worked in groups raise their hands, whereas those who worked alone do not.

NEW PARADIGM

Collaboration and *cooperation* represent a major new paradigm in business and industry, replacing that of *competition* which began with the Industrial Revolution and held sway well into the twentieth century.

Collaborative learning is consistent with modern engineering management practice and with what industry representatives tell us they want in our engineering graduates. *Competition* and *individual achievement* are outdated notions, and rightly so. W. Edwards Deming, father of the Total Quality Management (TQM) movement to use statistical methods to improve product quality, makes a compelling case [8]:

"We have grown up in a climate of competition between people, teams, departments, divisions, pupils, schools, universities. We have been taught by economists that competition will solve our problems. Actually, competition, we see now, is destructive. It would be better if everyone would work together as a system, with the aim for everybody to win. What we need is cooperation and transformation to a new style of management . . . Competition leads to loss. People pulling in opposite directions on a rope only exhaust themselves. They go nowhere. What we need is cooperation. Every example of cooperation is one of benefit and gains to them that cooperate. Cooperation is especially productive in a system well managed."

I hope you will seek opportunities for cooperation and collaboration with your fellow students, and that in doing so you will reap much greater rewards than you would have through competition and individual effort.

Conclusion

On a concluding note, I want to stress that implementing the success strategies presented in this chapter requires you to change – change how you think about things (your attitudes) and change how you go about these things (your behaviors). Thus, the value of any of the strategies mentioned above – indeed, the value of this entire book – depends on the extent to which you can make such changes. To help you succeed in what can be a difficult process, in Chapter 6 you will learn about the psychology of change, along with ways to gain insights into yourself, and a detailed process for your personal growth and development will be presented.

SUMMARY

This chapter addressed the important topic of "Making the Learning Process Work for You." Effective learning involves many skills and, as in developing any skill, practice makes perfect.

We began the chapter by discussing two very important learning skills: reading for comprehension and analytical problem solving. Your success in engineering study will depend in large part on your skill in these two areas.

We then described the process of organizing your learning process. We emphasized perhaps the most important academic success strategy in the chapter, if not in the entire book – the need to keep up in your classes by mastering the material presented in each class before the next class comes. Approaches for mastering material, along with important time and priority management skills, were also presented.

Next we discussed strategies for preparing for and taking tests. This subject deserves particular attention since most of your grade in math/science/engineering coursework will be based on your performance on tests and exams.

The chapter concluded with approaches for making effective use of one of the most important resources available to you – your fellow students. By working collaboratively with your peers, particularly in informal study groups, sharing information with them and developing habits of mutual support, you will learn more and enjoy your education more. At the same time, you will become well-prepared for the engineering work world, where teamwork and cooperation are highly valued.

REFERENCES

1. Polya, George, *How to Solve It: A New Aspect of Mathematical Method*, Princeton University Press, 2004.

2. Emmett, Rita, *The Procrastinator's Handbook: Mastering the Art of Doing It Now,* Walter & Company, New York, 2000.

3. Burke, Jane B., *Procrastination: Why You Do It, What to Do About It Now*, Da Capo Lifelong Books, 2008.

4. Knaus, William, *End Procrastination Now!: Get it Done with a Proven Psychological Approach,* McGraw Hill, 2010.

5. Covey, Stephen R., *The Seven Habits of Highly Effective People*, Simon & Schuster, New York, 1989.

6. Wood, J., "Sacrificing Sleep to Study Can Backfire," *Psych Central*. Retrieved on October 17, 2012, from www.psychcentral.com/news/2012/08/23/sacrificing-sleep-to-study-can-backfire/43520.html.

7. Smith, Karl A., "Cooperation in the College Classroom," Notes prepared by Karl A. Smith, Department of Civil Engineering, University of Minnesota, 1993.

8. Deming, W. Edwards, *The New Economics for Industry, Government, Education*, MIT Center for Advanced Study, Cambridge, MA, 1993.

PROBLEMS

1. Prepare a ten-minute talk on the methodology presented in Section 5.1 on *Reading for Comprehension*. Persuade a classmate from one of your other classes (not your *Introduction to Engineering* course) to listen to your talk and give you feedback on how well you understand and communicate the concepts.

2. Make a commitment to follow the steps in the methodology presented on *Reading for Comprehension* in Section 5.1 for one week. Be particularly attentive to the three "Before You Read" steps and to the recitation step "After You Read." Write a one-page paper on how the methodology impacted your learning.

3. Solve this problem following each of the four steps for analytical problem solving presented in Section 5.1. Use the problem-solving strategies of *solving a simpler problem* and *drawing a diagram*.

    A painter built a ladder using 18 rungs. The rungs on the ladder were 5.7 inches apart and 1.1 inches thick. What is the distance from the bottom of the lowest rung to the top of the highest rung?

4. Solve this problem following each of the four steps for analytical problem solving presented in Section 5.1 using the *guess and check* strategy. Using the *make a table* strategy. Using the *solve an equation* strategy.

    Amanda has 26 nickels and dimes in her piggy bank. The number of nickels is two fewer than three times the number of dimes. How much money does she have in her piggy bank?

5. Using the form presented at the end of this chapter, schedule your study time for one week. Attempt to follow the schedule. Write a one-page paper describing what happened.

6. Make a "To Do" list of things you need to do. Place each item in one of the four quadrants of Covey's matrix shown on Page 148. How many items are in Quadrant I? How many are in Quadrant II? How many are in Quadrant III and IV?

7. Go to your campus library and find a book on study skills. Check out the book and scan its "Table of Contents." Identify one interesting section and read it thoroughly. Then write an essay on why you picked the topic you did and what you learned about it.

8. If you studied for 100 hours, how many of those hours would be spent studying alone and how many would be spent studying with at least one other student?

9. If your answer to Problem 8 was that you spend most of your time studying alone, seek out a study partner in one of your math/science/engineering classes. Get together for a study session. Write down what worked well and what didn't work well.

10. Interview two junior or senior engineering majors and ask the following questions:

   a. What was the main difference they found between high school and university-level engineering study?

   b. What were the most important study skills they had to learn?

   c. What approach do they use to manage their time effectively?

   d. What do they think about group study?

	MONDAY	TUESDAY	WEDNESDAY	THURSDAY	FRIDAY	SATURDAY	SUNDAY
8-9							
9-10							
10-11							
11-12							
12-1							
1-2							
2-3							
3-4							
4-5							
5-6							
6-7							
7-8							
8-9							
9-10							

CHAPTER 6
Personal Growth and Student Development

*You will either step forward into growth,
or you will step back into safety.*

— Abraham Maslow

INTRODUCTION

The focus of this chapter is your personal growth and your development as a student. Your success as a student and, later, as an engineering professional will depend on the extent you grow and develop both during and after college.

We begin the chapter by tapping into the paradigm of "continuous improvement" espoused by U.S. business and industry, and we urge you to adopt a personal plan of continuous improvement for every area you need to strengthen or change. We call this process *student development*.

To achieve the changes that your student development plan will entail, we present a step-by-step process based on behavior modification theory.

To change yourself, you must first understand yourself. We therefore address three topics that conduce to self-understanding:

- The noted American psychologist Abraham Maslow's "Hierarchy of Needs" lists basic human needs that must be met before you can concentrate on achieving your highest need – *self-actualization*.

- A discussion of self-esteem shows you the domino effect that your personal development plan will have on your sense of self. As you make changes in thoughts and actions that move you closer to your goals, you will feel increasingly better about yourself.

- A presentation about peoples' personality types sharpens your understanding of yourself and others.

We then go into detail on the important topic of understanding others – which in large measure is an extension of the process of understanding yourself. Your success in your career will depend on your ability to work with people who are different from you, whether it be their personality type, learning style, value system, or culture.

Next, we return to and expand upon personal assessment. We home in on ways to identify your strengths and weaknesses so that your personal development plan is as comprehensive as possible. Within this context, we address two important (and often overlooked) areas of personal development: communication skills and mental/physical wellness.

Teamwork and leadership are then discussed as key parts of developing your communication skills, particularly interpersonal communications. Becoming effective as both a team member and leader will have a big impact on your success as an engineering student and, later, as an engineering professional.

Finally, we leave you with several motivational messages and point you to a large catalog of motivational and inspirational quotes that can help in strengthening your commitment to ongoing self-development.

6.1 PERSONAL DEVELOPMENT - RECEPTIVENESS TO CHANGE

I usually start the first meeting of my *Introduction to Engineering* course by asking:

"How many of you want to change something about yourself?"

Generally, only three or four out of 30 students raise their hands.

This resistance to change, I think, has the same roots as students' reluctance to seek help (discussed in Chapter 3) in that both are seen by students as an admission that something is "wrong" with them.

This, we have already shown, is a counterproductive attitude, but not for students alone. Resistance to change was a powerful force in post-World War II business and industry practices in the United States and a big factor in losing our number-one position in world manufacturing at the time. While other countries such as Japan, Korea, Taiwan, and Germany were striving for continuous improvement, we in the U.S. were satisfied with the *status quo*. Our motto for a long time was:

> *If it ain't broke, don't fix it.*

Only recently, in the last 40 years or so, has U.S. industry changed its tune dramatically, replacing the status quo paradigm with one of continuous improvement to regain its competitive edge.

TOTAL QUALITY MANAGEMENT

The term that has become synonymous with continuous improvement in business and industry practices worldwide is "Total Quality Management," or simply TQM. Developed in the 1950s and early 1960s primarily by Japanese industrialists – along with input from noted American statistician W. Edwards Deming – TQM espouses the philosophy that no matter how good we are, we should strive to continuously improve our quality [1]. Practitioners of TQM do not attach shame or feel resistance to change. They are not only receptive to change; they actively seek it out.

As part of your engineering education, you will undoubtedly hear a lot about TQM (or its current counterpart *Six Sigma*). For now, though, a general overview of the TQM process will suffice. Like the engineering design process, the TQM process consists of a series of steps or stages. The first involves defining what quality means – a definition that will change from one context to the next, depending on who the customer is and what the customer's needs are. Next, performance measures (*metrics*) are established. Last, a detailed plan is drawn up and implemented to meet or exceed the performance measures.

"PERSONAL" TOTAL QUALITY MANAGEMENT

I hope I can persuade you to adopt a ***personal TQM philosophy***. The "customer" can be you or someone else, such as your parents, your spouse or partner, your professors, or your future employer. Regardless of whom you choose, what's important is that you strive to change, grow, and improve yourself *continuously* in every area that impacts your effectiveness in meeting and exceeding the needs and expectations of your customer. Your motto should be:

> ***Even if it ain't broke, try to improve it.***

Consider, for example, a major league baseball player whose batting average is .315. This person was a superstar in high school, a superstar in college, and now a star in the major leagues. He makes $20 million a year. Yet he still works two hours a day with his batting coach to raise his average to .320. In fact, the reason his batting average is .315 is that he wasn't satisfied when it was .295.

The basic message is that successful people recognize the need to strive continuously to change, grow, and improve. Wanting to improve has nothing to do with the idea that there is something wrong with you.

REFLECTION

Reflect on the idea of growing, changing, and improving. Do you embrace the idea? What would you like to change about yourself? Become a better writer? Improve your "people skills"? Stop procrastinating? Learn to cook? Feel more comfortable in your own skin?

STUDENT DEVELOPMENT

When I talk about a personal TQM philosophy as applied to becoming a more effective student, I call it student development, and I tailor both the TQM process and vocabulary to make them more student-oriented. The general concept of continuous improvement, however, remains unchanged.

The cornerstone of student development is well-defined goals – the counterpart of TQM's first step of defining a customer's needs. As a student, your immediate goal is earning your B.S. degree in engineering. That's a given. You may have other goals as well – achieving a certain grade point average, finding the job you want upon graduation, and performing well in that job. Over the long term, you may want to have a successful career as a practicing engineer or become president of your own company.

In previous chapters, we have talked extensively about the importance of goals. With regard to student development, they play other roles as we;;: They specify the areas in which you need to grow, change, or develop; and they provide the necessary foundation for tracking the progress of your personal growth (the counterpart of TQM's second step of defining performance metrics). Only with clear goals can you, or anyone else, make value judgments about your behavior. Let me give you an example.

> If a student were to tell me, "My friends have invited me to go to Las Vegas this weekend," I would have no way of assigning a value judgment to her statement. If, however, she added, "I probably won't go, because I really want to graduate this year, and I'm a little behind in my thermodynamics class," a value judgment would be easy to make, and I would support her decision not to go.
>
> On the other hand, if the same student explained, "I just finished my final exams. I studied really hard and did great," a value judgment would again be easy and I'd be inclined to say, "Congratulations! You've earned it. Have a great time in Vegas."

In this example, it was not until I knew the student's goal *("I really want to graduate this year.")* that I could place a value judgment on her proposed weekend trip. Our goals, then, provide the context we need to assess what we do – or propose to do.

VALUE JUDGMENTS APPLIED TO OUR ACTIONS, THOUGHTS, AND FEELINGS

In forthcoming sections, we will examine in depth how our goals and value judgments fit into a larger process of change and personal growth. As a preface to this examination, let's look more closely at value judgments – the *metrics* of our personal TQM program – and the behaviors to which they apply.

In the example I just gave, my value judgments pertained to the student's action (or proposed action). **Actions**, what you say and do, are one part of human behavior. You also have **thoughts**, the ideas, attitudes, values, worldviews, and mindsets you hold; and **feelings**, the emotions you have. Obviously, your actions, thoughts, and feelings are deeply interrelated: Your feelings can affect your thoughts and actions, your actions can impact your thoughts and feelings, and so forth. But it is helpful to separate them out when talking about student development, for doing so establishes a framework for analyzing, understanding, and changing yourself.

Value judgments can be made based on an analysis of your actions, thoughts, and feelings. For this purpose, you can classify your actions in one of two ways: **productive** or **non-productive**.

<u>**Productive actions**</u> support the achievement of your goals.

<u>**Non-productive actions**</u> do the opposite: they interfere with or work against the achievement of your goals.

Similarly, you can classify your thoughts as either **positive** or **negative**.

<u>**Positive thoughts**</u> result in your choosing productive actions.

<u>**Negative thoughts**</u> result in your choosing non-productive actions.

If you return to the example on the previous page, you can see how thoughts can affect actions in either positive or negative ways. When the student in the example announced, *"My friends have invited me to go to Las Vegas for the weekend,"* she proposed an **action**. Before acting on the proposal, however, she voiced two **positive thoughts**: *"I really want to graduate this year"* and *"I probably won't go,"* which led to a **productive action** – i.e., staying home to catch up in her thermodynamics class.

In this case, we see positive thoughts bringing about a productive action. The reverse can also happen, as the following examples of negative thoughts leading to non-productive actions show:

Negative Thought	Non-Productive Action
"I'm so far behind, I don't get anything out of going to class."	Cut class.
"I learn better studying by myself."	Spend 100 percent of study time alone.
"Physics is too hard. I just can't do it."	Procrastinate; put off studying.
"Professors don't seem to want to help me. They make me feel stupid."	Avoid seeking help from professors outside of class.
"I don't like having my life run by a schedule."	Waste time by not scheduling it.
"I don't have time for student organizations."	Avoid participation in student organizations.
"I'm not good at writing and don't like doing it."	Avoid opportunities to develop writing skills.

Finally, our **feelings** can be classified as either **positive** or **negative**.

<u>**Positive feelings**</u> produce positive thoughts, which in turn lead to productive actions.

<u>**Negative feelings**</u> produce negative thoughts, which lead to non-productive actions.

Sources of feelings, particularly negative ones, are not always easy to pinpoint. Because they are often connected to our self-esteem, they may be hidden, locked away in our unconscious minds as nature's way of protecting us from dangerous or unpleasant experiences. If this is the case, it normally takes time and a concerted effort with the help of a therapist or counselor to uncover them.

Self-esteem and the consequent feelings we have, conscious or unconscious, are extremely important factors in the student development process. We will discuss these factors later in the chapter, but for now it is enough for you to know that feelings constitute a distinct part of being human. And, like our thoughts, they can be judged as either positive or negative.

To give you just one example of how our feelings can affect our thoughts and actions, read the following story about "Jane R."

> Jane R. gets terrible feelings of anxiety when she has to speak in public. She has thoughts like, "I'm a lousy speaker." Before she speaks, she gets so nervous that she does a poor job.
>
> Out of desperation, she goes for counseling. During therapy, she recalls that in elementary school she was criticized by her teacher when called on to read aloud to the class. This experience left her traumatized.
>
> In therapy she realizes that it was okay for a third-grader to make a mistake when reading aloud. She also realizes that her teacher wasn't intentionally trying to harm her. As a result, she is able to forgive herself and her teacher. By diffusing the negative feelings, she begins to think, "Maybe I can do a good job of speaking in public." And she

If we were to analyze the cause-effect relationships of Jane R.'s feelings, thoughts, and actions first before and then after her therapy, the process could be diagrammed as follows:

BEFORE THERAPY

Negative Feeling	→	Negative Thought	→	Non-Productive Action
Very anxious prior to making presentations		"I'm a lousy public speaker."		Does a poor job when making a presentation

AFTER THERAPY

Positive Feeling	→	Positive Thought	→	Productive Action
Feel okay, even optimistic about future presentations		"Maybe I can do a good job of speaking in public."		Performs well on subsequent presentations

REFLECTION

Reflect on your recent actions. What are examples of actions you took that were supportive of your goal of becoming an engineer (productive actions)? What thoughts led you to take these actions (positive thoughts)? What are examples of actions you took that worked against or interfered with your goal of becoming an engineer (non-productive actions)? What thoughts led you to take those actions (negative thoughts)?

THERAPY AND COUNSELING AS CHANGE AGENTS

Jane R.'s story calls attention to the potential value of counseling and therapy in the process of personal development. Although psychotherapy typically requires a costly and lengthy commitment, some counseling may be available to you on your campus through the student health center or counseling center.

Jane R.'s case also illustrates the basic premise of counseling or therapy, which is that any ongoing, unresolved negative feelings likely resulted from some traumatic childhood experience, which has been buried away in the unconscious mind. If this is the case, the job of the psychotherapist is to dig beneath one's surface feelings to find their root cause. Once uprooted, the originating source and feelings generally lose the cathectic force they were able to exert over the years, and the individual is then able to deal with them on a conscious, rational level – or as Jane R. did, discard them entirely.

This is, of course, a very cursory overview of the counseling process, but I hope it gives you a basic understanding of how therapy works and when it is useful to overcome certain obstacles standing in the way of your personal growth.

BEHAVIOR MODIFICATION AS A PROCESS FOR CHANGE

Behavior modification is another effective mechanism for changing your actions, thoughts, and feelings. It's probably the most accessible and practical approach if you are truly committed to personal growth.

The premise of behavior modification is somewhat different from that of counseling/therapy. It assumes that human behavioral change should start with your actions, which you consciously choose. If these actions are non-productive, you have the option to replace them with productive actions and so instigate a process that filters down to your thoughts and feelings.

According to behavior modification theory, you have less control over your thoughts than your actions. You cannot help having negative thoughts. However, you can become more conscious of them and thus have the power to change them. Generally, you can do this by finding a higher context for your thinking. Your goal will provide that context.

An Example

You have a test coming up in your math class. A productive action would be to study from 7:00 to 10:00 p.m. tonight. Behavior modification would hold that you are completely capable of choosing that action. However, you may have thoughts such as "I'd rather go out with my friends" or "I'm tired and don't feel like studying tonight." These are negative thoughts because, if acted on, they will lead you to a non-productive action (i.e., not studying).

Your challenge is to recognize that such thoughts are negative and try to change them. For example, the thought that "I don't feel like studying tonight" can be changed to "I really do want to study tonight because doing well in my math class will move me closer to getting my engineering degree."

As noted earlier, you have much less direct control over your feelings, because they are often tied to your self-esteem. While the issue of self-esteem will be dealt with in a subsequent section, you saw roughly how counseling/therapy starts with one's feelings to enact changes in one's behaviors. A similar rough sketch can be drawn to show how behavior modification works:

If you begin to choose productive actions in support of a personal goal, and if you work to change negative thoughts to positive ones in support of those actions, in time you will feel more positive about yourself and about your life.

STUDENT SUCCESS MODEL

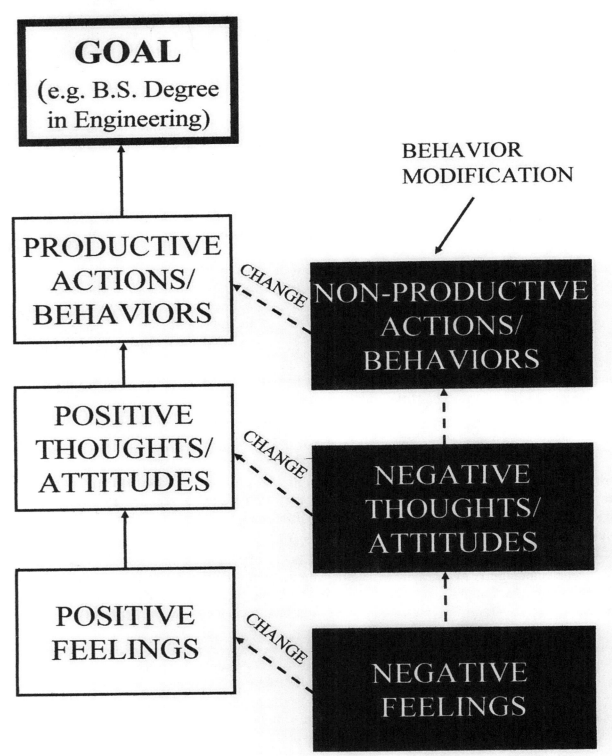

As illustrated in the Student Success Model on the previous page, achieving your goal of graduating in engineering will likely require that you change your actions from non-productive to productive ones, your thoughts from negative to positive, and your feelings from negative to positive. Through these changes you will grow and develop as a student.

6.2 MAKING BEHAVIOR MODIFICATION WORK FOR YOU: THREE STEPS TO ACHIEVE CHANGE

Sound easy? Just change those negative thoughts to positive ones, start choosing productive actions, and everything will be just great. Actually, you will find that change is not so easy. Changing your behavior requires you to successfully navigate three steps, each of which can present significant barriers to change:

| Step 1. **Knowledge** - "You know what to do." |
| Step 2. **Commitment** - "You want to do it." |
| Step 3. **Implementation** - "You do it." |

STEP 1. KNOWLEDGE - "YOU KNOW WHAT TO DO."

By *knowledge*, we mean: ***You know what to do.***

One of the main purposes of this book is to provide you with the knowledge of those strategies and approaches that will enhance your effectiveness as an engineering student. Much of that knowledge was presented in Chapters 3, 4, and 5. Hopefully, you have studied those chapters and recognize that implementing many of the strategies and approaches presented there will require you to change your behaviors.

Some examples of these changes are presented below.

Non-Productive Action	**Productive Action**
Neglect studying	CHANGE ⟶ Devote significant time and energy to studying.
Delay studying until test is announced	⟶ Master the material presented in each class prior to the next class.
Study 100% alone	⟶ Spend time studying with other students.
Come to lectures unprepared	⟶ Review notes, read text, and attempt problems prior to each lecture.
Avoid professors	⟶ Interact regularly with professors outside the classroom.
Spend little time on campus	⟶ Immerse yourself in the academic environment of the institution.
Avoid participation in student organizations	⟶ Participate actively in student organizations.

Gaining the knowledge presented in Chapters 3, 4, and 5, however, does not guarantee that you put it into practice.

New Knowledge Is No Guarantee

A prime example of how new knowledge does not always produce change is smoking cigarettes. When I was growing up, people simply did not know that smoking caused cancer. Today, no one can deny knowing that it does. The relationship between smoking and lung cancer is an example of a new knowledge base. However, many people still smoke. They failed to change because they did not make a commitment to act on the new knowledge.

STEP 2. COMMITMENT - "YOU WANT TO DO IT."

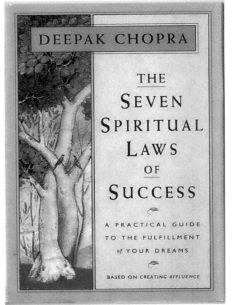

By *commitment,* we mean that you not only know what to do but that: ***You want to do it.***

Developing that commitment requires you to go through the process of examining each academic success strategy and deciding whether you want to put it into practice. Do I want to schedule my time? Do I want to study from class to class rather than from test to test? Do I want to prepare for each lecture? Do I want to study with other students? Do I want to make effective use of my professors? Do I want to spend more time on campus? Do I believe these strategies and approaches will enhance my academic success?

A commitment to doing something typically relies first and foremost on a change in attitudes. That is, to commit to your new knowledge base, you need to first change any pre-existing negative attitudes. To do this, you need to be aware of those attitudes, particularly the ones that will obstruct your growth. A quote from Deepak Chopra's excellent book *Seven Spiritual Laws of Success* [2] shows the way:

Most of us, as a result of conditioning, have repetitive and predictable responses to the stimuli in our environment. Our reactions seem to be automatically triggered by people and circumstances, and we forget that these are still choices that we are making in every moment of our existence. We are simply making these choices unconsciously.

If you step back for a moment and witness the choices you are making as you make those choices, then in just this act of witnessing, you take the whole process from the unconscious realm into the conscious realm. This procedure of conscious choice-making is very empowering.

If negative attitudes are keeping you from changing your behaviors, try the following approach:

(1) Identify key areas in which your attitudes (e.g., studying with other students, seeking one-on-one instruction from your professors during office hours, becoming involved in student organizations) have the potential to significantly impact your academic success.

(2) For each area, describe those attitudes.

(3) For each attitude, answer the question, "Is this attitude working for me or against me?"

(4) For each negative attitude, try to figure out its origin (i.e., "Where did it come from?").

(5) Then, answer the question, "How can I change this negative attitude to one that will work for me?"

As this example illustrates, changing your attitudes is a necessary precursor to changing your actions.

Example of This Process

Let's imagine that a student believes he is failing a math course because the professor comes across as boring, unprepared, and aloof. This student has the attitude that "I can't pass a course if I don't like the professor." He sees himself as a "victim" in which passing his math course is totally in the control of his professor. Once he becomes conscious that this is a negative attitude (interfering with his goal of academic success) and realizes that the attitude can be changed, he can go about changing it to a positive one.

An example of a positive attitude would be: "I can pass the course even if I don't like the professor, but it is going to require me to pursue strategies such as sitting in on another instructor's lecture, getting old exams, or seeking help from students who passed the course previously."

Freshman Engineering Student Attitudes Survey. At this point, you might find it useful to get some feedback on the attitudes you hold. Go to: www.discovery-press.com/discovery-press/studyengr/attitudesurvey.pdf. There you will find an instrument you can use to measure your attitudes in 70 areas related to your success as an engineering student. This instrument is based on the Pittsburgh Freshman Engineering Attitudes Survey (PFEAS) [3]. If you identify any negative attitudes, try to change them using the steps described in the previous section.

Lifelong Learning. One very important attitude is how you view learning and particularly *lifelong learning*. Do you think learning is confined to the time you are in school and to the courses you are required to take? If so, this is a limited view that will not put you in good stead in your life and career.

Developing a view toward lifelong learning is particularly important in engineering because of the rapid growth of technological knowledge and innovation. Depending on your specialty, the half-life of knowledge in your field – the time it would take for half of everything you learn in school and beyond to become obsolete – ranges from 2.5 years to 7.5 years. This means that as a practicing engineer you will have to update one-half of everything you know every few years.

Given this rate of change, it is no surprise that one of the eleven attributes ABET requires of engineering graduates is:

a recognition of the need for, and an ability to engage in, life-long learning

If you think you'll know everything when you march up the aisle to receive your bachelor of science degree in engineering, you couldn't be more wrong. Industry representatives often call new engineering graduates "freshouts," a description referring to the fact that it will be several years before they will know enough to even earn their salary.

To a great extent, the process of expanding your knowledge and skills and keeping up-to-date in your field will be your responsibility. Take every opportunity to read books and journals. Talk to more knowledgeable people where you work. Attend professional meetings. Take short courses and attend seminars. Take additional university courses, whether as part of an advanced degree program or for your own professional development.

REFLECTION

How do you view learning? Do you limit your learning mainly to what is required in your classes? Or do you look for other opportunities to learn? What are some examples of things you could do to learn more than what is covered in your classes? What is your view toward lifelong learning? Are you excited about learning continuously through your career and life?

STEP 3. IMPLEMENTATION - "YOU DO IT!"

The final stage is implementation. By *implementation*, we mean that you not only know what to do and want to do it, but that: ***You do it!***

This is probably the most difficult step to take. Change can be hard, no matter how knowledgeable and committed you may be.

There are many reasons why you may fail to change behaviors that you know you should – behaviors that, in effect, work against you. For example, let's assume that you are not putting sufficient effort into studying and are doing poorly in your classes. How can this be? You want to be successful in school. You want to get all of the rewards that a career in engineering will bring to you. Still, you are not doing what is required for success.

BARRIERS TO IMPLEMENTING PRODUCTIVE ACTIONS

One primary reason you may not be willing to change is there is payoff for you to keep doing what you are doing. You have adopted your current behavior patterns because they satisfy some need or want. Abandoning old, familiar behaviors means, to some extent, finding new ways to satisfy those needs.

For example, you may go home or to your residence hall as soon as you get out of class. You may do this because you get a great deal of pleasure from the distractions you find there, such as friends, family, TV, music, food in the refrigerator, telephone, and the Internet. Choosing to stay on campus to study with other students, to seek help from professors, or to use the resources of the library may be less enjoyable to you. You may have to really work on yourself to change your thought from "I enjoy going home" to "If I go home, it's likely that I won't study. I'm going to stay at school until I get my work done."

Or you may have difficulty choosing to do things you don't find easy or pleasurable. Most people do. The impact on one's success of choosing to do difficult tasks rather than enjoyable

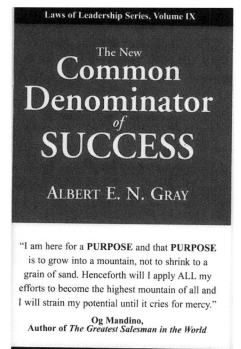

ones is perhaps best stated in a remarkable presentation by Albert E. N. Gray to a national meeting of life insurance underwriters in 1940 titled "The Common Denominator of Success" [4]

As so well put by Mr. Gray:

The common denominator of success – the secret of every [person] who has ever been successful – lies in the fact that they formed the habit of doing things that failures don't like to do.

Gray goes on to explain why a person would choose to do things they don't like doing:

Successful [people] are influenced by the desire for pleasing results. [They] have a purpose strong enough to make them form the habit of doing things they don't like to do in order to accomplish the purpose they want to accomplish. Failures are influenced by the desire for pleasing methods and are inclined to be satisfied with such results as can be obtained by doing things they like to do.

I hope your commitment to your goal of becoming an engineer is strong enough to influence you to choose productive behaviors, even if you don't like doing some of them.

I would also encourage you to read the complete text of Mr. Gray's inspiring presentation. See: www.discovery-press.com/discovery-press/studyengr/CommonDenominator.pdf.

Personal Perspective

In some ways I don't envy you. You have so many more distractions than I had when I was a student. When I was a student there were no cell phones, video games, MP3 players, PCs, Internet, YouTube, email, text messaging, or 500 channels of high definition television. When I look at the list of the "50 Greatest Technological Inventions of the Past 25 Years" in Chapter 2 (Page 71-72), I see lots of "pleasing methods" that I didn't have to tempt me.

To a great extent, your ability to manage those things so that they help you achieve your goals rather than interfere with you achieving your goals will be a key factor in your success. I hope you are able to make the right choices. I don't know of anyone who says, "I sure wish I had watched more television and played more video games when I was younger."

REFLECTION

Reflect on Albert E.N. Gray's concept that people who are successful develop the habit of doing things that people who are failures don't like to do. Do you see the wisdom in this idea? What are things that you don't like to do that need to be done if you are to be successful in your math/science/engineering courses? Is your commitment to your goal of graduating in engineering strong enough to influence you to develop the habit of doing these things?

There are other reasons why you might choose non-productive behaviors. Human psychology is very complicated and doesn't always make sense. You may be afraid to study because if you do and still fail, it will reflect on your ability. Or you may be trapped in a "victim" role, preferring to blame your failure on external factors or other people. Perhaps you feel you were forced to go to college by your parents. By not studying you are showing them that you are your own person – that you are not going to do what they want you to do. **Making change requires you to accept responsibility for your actions and to begin to view yourself as the creator of your life.**

6.3 UNDERSTANDING YOURSELF

There is one subject that you don't study in school – a subject that is the key to your happiness and quality of life. That subject is YOU! The most exciting adventure you will ever embark on is the journey of self-discovery and self-awareness.

Understanding yourself is an essential aspect of becoming a productive and happy person. There are other benefits as well. As you grow in your understanding of yourself, your ability to understand other people will likewise grow.

But nobody said it would be easy. As Benjamin Franklin noted:

There are three things extremely hard: steel, a diamond, and to know one's self.

Understanding yourself is a lifelong process because human beings are very complex. As a result there are many different models or frameworks to describe human behavior and psychology.

Some of these frameworks are more useful than others. In fact, some are not even valid. I always wonder how I am supposed to believe that one-twelfth of the people in the world (all those born under my zodiac sign) have something in common with me.

Others are over-generalizations. For example, I have a friend who read a book on the importance of birth order [5]. What a bore he became! He went around asking people their birth order. If they said they were first-born, he would tell them that they were reliable, conscientious, driven to succeed, serious, self-reliant, well-organized, and on and on. This is an example of a framework that tries to put all people into one of three categories (i.e., firstborn, middle child, or last born) – an obvious oversimplification.

In Chapter 3 we presented a model for the different ways students prefer to receive and process new knowledge. Hopefully, by understanding your preferences, you have grown and developed in your ability to better design your learning process.

In this section, we will present two additional models that can be useful to you as an engineering student. The first is **Maslow's Hierarchy of Needs** [6]. This hierarchy will give you an understanding of basic needs that must be met if you are to be motivated to succeed in your studies.

One of these needs is feeling good about yourself. Because your *self-esteem* is a very important factor in your productivity and overall happiness, we will address it in some detail.

Finally, we will discuss the **Myers-Briggs Type Indicator** as a way to characterize different personality types, and we will note how engineering students on average differ in personality types from the general population.

MASLOW'S HIERARCHY OF NEEDS

Famed psychologist Abraham Maslow focused his research on the relationship between human motivation and needs, particularly unmet needs. Motivation, he claimed, is the inner drive that propels behaviors and actions. To illustrate the connection between motivation and needs, he developed this hierarchy:

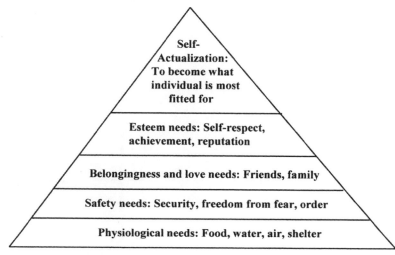

MASLOW'S HIERARCHY OF NEEDS

According to Maslow's theory, needs must be satisfied from the bottom up. If a lower-level need exists, you will be highly motivated to satisfy that need. When lower-level needs are satisfied, higher-level ones become important, and you become motivated to satisfy those needs.

At the lowest level are **your physiological needs** for food, water, air, and shelter. If you are

---

Needs Are Not Wants

It is important to distinguish between needs and wants. **Needs** are things that you must have, things that are essential. **Wants** are things that you desire. For example, you may want to have an expensive car, but having one is not essential. Don't let unnecessary wants distract you from academic success.

---

reading this book, you probably are satisfying these needs. If not, it is unlikely that you will be able to focus on your academic work.

At the second level are **your safety needs**, including the need for security and freedom from fear of physical and psychological threats. Again, I hope that these needs are not issues for you. If you are afraid of a bully in your residence hall or a former boyfriend or girlfriend, it is doubtful that you will be able to concentrate adequately on your studies.

At the third level are **your social needs**, such as needing to belong, to be accepted, and to receive support and affection from others. These social needs are generally met by family and friends. If you left home to go to college, you may be experiencing a period in which these social needs are not being met. It is important, therefore, for you to develop new friends and relationships in your new environment. Unmet social needs can interfere with your studies. The good news is that many of your classmates are also looking to satisfy these same needs.

Abraham Maslow

Maslow's fourth level centers on **your needs for esteem**, including self-respect, achievement, and recognition. You need to feel good about yourself and to feel as though you have the respect and appreciation of others. As an engineering student, you build self-esteem through appreciation for your academic success and through co-curricular accomplishments. But equally important is how you treat other people. (We will elaborate on the topic of self-esteem in the next section.)

The fifth and highest level is **your need for self-actualization**. Self-actualization means fully developing your abilities and ambitions. It is the need you have to reach your highest potential, or put in simple terms, "to do your best." This is the need that causes you to want to excel on a test, to do your best in a game of tennis, and generally to learn, to grow, and to develop in a myriad of ways. As Maslow explains:

> Even if all other needs are satisfied, we may still often expect that a new discontent and restlessness will soon develop, unless the individual is doing what he or she, individually, is fitted for. Musicians must make music, artists must paint, poets must write if they are to ultimately be at peace with themselves. What humans can be, they must be. They must be true to their own nature. This need we may call **_self-actualization_**.

Obviously, to be a successful student, you must be able to pursue your need for self-actualization. This means that you must first satisfy your physiological, safety, social, and esteem needs.

SATISFYING YOUR NEED FOR SELF-ESTEEM

As indicated by Maslow's Hierarchy of Needs, self-esteem is a fundamental human need. We cannot be indifferent to the way we feel about ourselves. Self-esteem is a critically important factor in virtually every aspect of life. It influences what we choose to do, how we treat others, and whether we are happy or not.

Many problems faced by our society such as drug and alcohol abuse, crime and violence, poverty and welfare abuse, teenage pregnancy, the disintegration of the family, and the high dropout rate among high school students are directly related to the low self-esteem of many of our citizens. Several years ago, the California Legislature established a Task Force to Promote Self-Esteem to make recommendations on what the state could do to enhance the self-esteem of its citizens [7].

The Task Force defined *self-esteem* as:

> Appreciating my own worth and importance and having the character to be accountable for myself and to act responsibly toward others.

According to Nathaniel Branden [8], self-esteem is made up of two interrelated components:

Self-efficacy - your sense of competence

Self-respect - your sense of personal worth

To be self-efficacious is to feel capable of producing a desired result. Self-efficacy derives from your confidence in the functioning of your mind and your ability to understand, learn, and make decisions.

Self-respect comes from feeling positive about your right to be happy, from feeling that you are worthy of the rewards of your actions, and from feeling that you deserve the respect of others.

It is important to have <u>both</u> self-efficacy and self-respect. If you feel competent but not worthy, you may accomplish a great deal, but you will lack the capacity to enjoy it. You may feel that you must continually prove your worth through achievement. Overachievers and "workaholics" are generally striving to meet their need for self-respect by feeling competent and productive.

There is a strong correlation between our self-esteem and our behaviors. According to Branden, healthy self-esteem correlates with:

Rationality	Realism
Intuitiveness	Creativity
Independence	Flexibility
Cooperativeness	Willingness to admit mistakes
Benevolence	Ability to manage change

Poor self-esteem correlates with:

Irrationality	Blindness to reality
Rigidity	Fear of the new and unfamiliar
Rebelliousness	Inappropriate conformity
Defensiveness	Overcontrolling behavior
Fear of others	Hostility toward others

REFLECTION

Consider the ten items listed in this section that correlate with healthy self-esteem. How many of the items describe you? Consider the ten items listed in this section that correlate with poor self-esteem. How many of those items describe you? What might you do to change yourself with regard to any items that correlate with poor self-esteem?

It is no surprise that high self-esteem is one of the best predictors of personal happiness [9]. The value of self-esteem is not merely that it allows you to feel better, but also that healthy self-esteem will be a key factor in your productivity and success. According to Branden:

> **High self-esteem** seeks the challenge and stimulation of worthwhile and demanding goals. Reaching such goals nurtures self-esteem. **Low self-esteem** seeks the safety of the familiar and the undemanding. Confining oneself to the familiar and the undemanding serves to weaken self-esteem.

How can you enhance your self-esteem? The feedback loop between your actions and your self-esteem described by Branden points the direction. That is, the level of your self-esteem influences how you act. Conversely, how you act influences the level of your self-esteem.

Recall our discussion of behavior modification in Section 6.2, which is based on the premise that you **choose** your actions. You can choose productive actions or you can choose non-productive ones. You have less control over your thoughts, but you can catch negative thoughts and work at changing them to positive thoughts. Behavior modification also asserts that if you choose productive actions in support of a personal goal and you work at changing your negative thoughts to positive thoughts in support of those actions, in time your feelings will change. You will feel better about yourself and your life. Your self-esteem will improve.

*Your college years provide a unique opportunity for you to enhance your self-esteem by building **both** your self-efficacy **and** your self-respect.*

Your engineering education will strengthen your problem-solving skills, your technical knowledge, and your ability to work with others. All of this will increase your confidence in your ability to face life's challenges and to achieve whatever goals you set for yourself. Through this process, you will build your self-efficacy.

Your engineering education will afford opportunities to build your self-respect and your feeling of personal worth as well. Academic success will bring positive feedback from your professors and fellow students. More tangible rewards such as scholarships, internships in industry, and admission to graduate school can be yours. You can be president of an engineering honor society, be the team leader of your institution's entry in a national engineering student competition, be paid to work on a professor's research project, or be the co-author of paper that is presented at an international conference. These accomplishments will be respected by others and will enhance your sense of self-worth. And you won't have to feel this way:

Copyright 2000 by Randy Glasbergen.

"I'VE GOTTA BE ME . . . BUT I CAN'T HELP THINKING SOMEONE ELSE WOULD BE MORE QUALIFIED!"

Success in engineering study will enhance your feeling of competence and your self-respect. These together will build healthy self-esteem. But it is up to you. You can let the negative feelings associated with low self-esteem produce negative thoughts that lead you to non-productive actions and failure. You can also choose productive actions and positive thoughts that will lead you to success and to feeling good about yourself.

MYERS-BRIGGS TYPE INDICATOR (MBTI)

Individuals are different. We each have preferences for how we interact with the world around us, how we learn, and how we make decisions.

I, for example, am an extrovert. I am energized by interactions with other people. I like lots going on around me and like to get tasks accomplished. I am a linear thinker. I can figure out anything that is logical. The hardest assignments I had in high school were to memorize poems. I just couldn't do it. It wasn't logical. Furthermore, I'm not very creative. I would be the worst at painting a picture or decorating an empty room.

Carl Jung

Finally, I like things to be planned and orderly. I don't like surprises. I'm not very spontaneous. When I go on vacation, I make reservations. My worst fear would be to arrive somewhere and find there were no accommodations.

You may be very different from the way I am. You may have very different preferences. People may tire you out. You may prefer to work alone. You may be very creative and artistic. You may prefer being flexible and spontaneous as opposed to being organized and orderly. You might like to arrive in a European city late at night with no reservations. You may be very intuitive, easily grasping the "big picture" through your imagination or bursts of insight.

The famous Swiss psychologist Carl Jung did seminal work on psychological types [10]. Jung's work led to the Myers-Briggs Type Indicator (MBTI), a matrix widely used today to lead to self-understanding [11]. The MBTI characterizes individuals in four areas:

(1) Does the person's interest flow mainly to:	
The outer world of actions, objects, and persons?	E-extrovert
The inner world of concepts and ideas?	I-introvert
(2) Does the person prefer to perceive:	
The immediate, real, practical facts of experience and life?	S-sensing
The possibilities, relationships, and meanings of experiences?	N-intuiting
(3) Does the person prefer to make judgments or decisions:	
Objectively, impersonally, considering causes of events and where decisions may lead?	T-thinking
Subjectively and personally, weighing values of choices and how they matter to others?	F-feeling

(4) Does the person prefer mostly to live:	
In a decisive, planned, and orderly way, aiming to regulate and control events?	J-Judging
In a spontaneous, flexible way, aiming to understand life and adapt to it?	P-Perceiving

The result of this characterization is to place individuals into 16 personality types based on combinations of four pairs of letters (E or I, S or N, T or F, J or P).

I would encourage you to learn not only about your own personality type but also about the other 15 personality types. Learning about other personality types will help you understand how others may differ from you.

There are two ways to find out your personality type. One is to take the MBTI test. It is very likely that it is administered somewhere on your campus – either by the testing office, the counseling center, or the career center.

The second way is to take a similar test that is based on the same research that led to the MBTI: the Keirsey Temperament Sorter II. The handy thing about this test is that it can be taken and scored free of charge on the Internet at: www.keirsey.com/sorter/register.aspx.

The Keirsey Temperament Sorter II requires you to choose one of two descriptors for each of 70 items. Based on the results, you will be placed into one of the same 16 categories as the MBTI. These are also described in terms of four temperaments, each having four variants, shown below with the corresponding Myers-Briggs personality types.

Temperament	Variant 1	Variant 2	Variant 3	Variant 4
Guardian	Supervisor (ESTJ)	Inspector (ISTJ)	Provider (ESFJ)	Protector (ISFJ)
Artisan	Promoter (ESTP)	Crafter (ISTP)	Performer (ESFP)	Composer (ISFP)
Idealist	Teacher (ENFJ)	Counselor (INFJ)	Champion (ENFP)	Healer (INFP)
Rational	Field Marshall (ENTJ)	Mastermind (INTJ)	Inventor (ENTP)	Architect (INTP)

One feature of the Keirsey website is that it enables you to access descriptions of each of the 16 personality types. If those descriptions aren't enough, I highly recommend two excellent resources: Keirsey's book *Please Understand Me II: Temperament, Character, Intelligence* [12] and *Do What You Are: Discover the Perfect Career for You Through the Secrets of Personality Type* [13].

You might be interested to know that engineering students tend to differ from the general population, as the table below shows [14]:

	General Population	Engineering Students
Introvert (I)	30%	67%
Intuiting (N)	30%	47%
Thinking (T)	50%	75%
Judging (J)	50%	61%

As you can see, engineering students have a greater tendency than the general population to be introverts rather than extroverts, to prefer to use a logical approach to make decisions, and to prefer to live in a planned, orderly way. Although a higher percentage of engineering students are intuitors than the general population, more than half of engineering students prefer to gain information using their senses. In Myers-Briggs lingo, the most frequent personality type found among engineering students is ISTJ, followed in order by ESTJ, INTJ, INTP, and ENTJ.

6.4 UNDERSTANDING OTHERS/RESPECTING DIFFERENCES

One of the most important areas of your personal growth is respecting people who are different from you. Engineering, now more than ever, is a team-oriented profession. Your success as both an engineering student and engineering professional will be closely related to your ability to interact effectively with others. As an engineer, you will be required to work with, supervise, and be supervised by people differing from you in learning styles and personality types and in gender, ethnicity, as well as cultural background.

DIFFERENCES IN LEARNING STYLES AND PERSONALITY TYPES

Understanding and respecting the different learning styles presented in Chapter 3 and the various personality types presented in Section 6.3 of the current chapter can help you work effectively with people who differ from you. Ned Herrmann, in his excellent book *The Creative Brain,* [15] puts forth some pertinent ideas about such differences:

(1) Differences are not only normal, but they are also positive and creative.

(2) Appreciating and using these mental differences make change easier to deal with because it makes us more creative.

(3) As we appreciate the full spectrum of mental gifts – ours and those of others – we can make better choices in our lives, especially in selecting educational and career directions.

(4) If those who manage others acknowledge and honor personal preferences and give people the chance to match their work with their preferences, they will be able to see tremendous gains in productivity.

(5) As we learn to value and, above all, <u>affirm</u> one another's unique mental gifts, we can participate in the formation of true community – perhaps our best hope for survival in this strife-torn world.

I'm sure you'll agree that this is an impressive list of the benefits that can come to you by better understanding the differences in people's learning styles and personality types.

ETHNIC AND GENDER DIFFERENCES

Other critically important areas of difference concern gender and ethnicity. Forty years ago, 98 percent of engineers in the U.S. were white males. This is no longer the case. The percentage of women and ethnic minorities among engineering and computer science graduates has been increasing steadily. As the chart below indicates, 43 percent of engineering graduates were women, ethnic minorities, and foreign nationals. And that percentage is expected to grow.

B.S. Degrees Awarded in Engineering and Computer Science
Ethnic and Gender Groups - 2010/11 [16]

	Number	Percentage
Non-Minority Women	8,504	11.1%
African-Americans	3,251	4.2%
Hispanics*	5,613	7.3%
Native-Americans	360	0.5%
Asian-Americans	9,541	12.4%
Foreign Nationals	5,597	7.3%
TOTAL	**32,866**	**42.8%**

* Excluding Puerto Rico

Unfortunately, prejudice, bigotry, and discrimination continue to be problems in our society. We tend to have a compulsive need to build ourselves up by putting others down.

How we treat others is closely related to self-esteem. If we don't feel good about ourselves, it is likely that we won't feel good about others. This is well stated in the report of the California Task Force to Promote Self-Esteem [7]:

> The more we appreciate our own worth and importance, the more we are able to recognize and appreciate the worth and importance of others as well.

We now live in a multi-ethnic, multicultural society. The old *melting pot* idea is no longer operative. As a nation, we are now more like a "mosaic," or a mixture of separate and different peoples, each having its own unique characteristics. Indeed, the concept of a *melting pot* has become offensive. Why should I strive to be the same as you? The quality of my life and the contributions I make are related to my special qualities and uniqueness as an individual. If I come from a different background and experience than you, I can act and think in ways that you cannot.

As an engineering student and a future engineering professional, you need to learn to respect and value people from different ethnic backgrounds. This may require you to work through certain prejudices, even if they seem as "innocuous" as telling racist jokes. Just remember a prejudice is nothing more than a thought or attitude that you have the capacity to change.

A Personal Story

I grew up in a very racist environment. At the time I completed high school in 1957 in Jacksonville, Florida, the worst forms of institutional racism and legalized segregation were in effect. Certainly with that kind of upbringing, I had a lot of changing to do. After spending several years at a northern liberal college, I became a zealous supporter of the civil rights movement and came to abhor the evils of the racist society I grew up in. This is undoubtedly why I devoted a significant portion of my professional career to working affirmatively to undo the effects of racism.

I could tell innumerable stories of how I became aware of the many negative impacts of racism. One that comes to mind is the lesson I learned about ethnic jokes. I used to have a favorite Polish joke, one I thought "won me points" whenever I told it. Then, when I was working on my Ph.D. degree, I took a Russian course to meet my language requirement. I arrived a bit late to the first class, only to find the instructor agitated and upset.

As it turned out, he was of Polish descent and the subject of Polish jokes had somehow come up. At the second class meeting, he brought in a research paper documenting the demeaning effects the Polish joke phenomenon had on the self-image of Polish-American children. I promise you that I have never told that Polish joke nor any other ethnic joke since that day.

As we discussed in Section 6.2 and as my personal story demonstrates, new knowledge can produce change. Sensitivities can develop by understanding the injustices around us.

For many years I visited universities regularly to assist them with their efforts to increase the representation of underrepresented minorities in their engineering programs. I always asked to meet with minority students to get their perspective. It broke my heart when they told me that white students resisted forming laboratory groups with them, left the seats next to them vacant, or acted surprised when they did well on tests. I hope you will not engage in these very harmful and unnecessary behaviors.

STEREOTYPING IS UNNECESSARY AND UNFAIR

One of the primary bases of prejudice is *stereotyping*.

A *stereotype* is a fixed conception of a person or a group that allows for no individuality.

Engineering students are often labeled as *nerds* who care only about how things work and have no interest or skill in dealing with people. The obvious problem with general descriptions like this is that the stereotype doesn't apply to all individuals in the group.

Just as I'm sure you would not like to be labeled as a *nerd* merely because you are an engineering student, I hope you will refrain from stereotyping others. The best way to approach people who differ from you in any way is to suspend judgment. Resist the urge to draw conclusions about someone you don't even know. Indeed, view occasions to get to know individuals as opportunities to learn!

Stereotyping also works against gender parity. Sheila Widnall, Secretary of the Air Force in the Clinton administration, wrote a powerful paper [17] outlining some of the obstacles experienced by women who study science and engineering. According to Dr. Widnall:

> Studies of objective evaluations of the potential and the accomplishments of women give quite discouraging results. Such studies in which male or female names are applied to resumes, proposals, and papers that are then evaluated by both male and female evaluators consistently show that the potential and accomplishments of women are undervalued by both men and women, relative to the same documents with a male attribution.

I hope that you agree that we have a long way to go in providing equal and fair treatment to <u>all</u> people.

Reflection

Reread Secretary Widnall's passage above. Is it clear what she is saying? Do you think <u>you</u> would have a different reaction to the same resume if it had a man's name on it than if it had a woman's name on it? Do you think you would be more inclined to fund a proposal if it was attributed to a man than if it was attributed to a woman? Do you think the unequal treatment described in the passage is widespread? How does the thought that it is make you feel?

YOUR EFFECTIVENESS IN CROSS-CULTURAL COMMUNICATIONS

What can you do to improve your effectiveness in working with and communicating with people who differ from you? First and foremost, you should seek out opportunities to interact with men and women from different ethnic and cultural backgrounds. You can learn a great deal from them and improve your interpersonal communication skills in the process. If you truly want to grow in this area, take a course in cross-cultural communications. I think you will not only find the subject very interesting, you also will learn skills that will come into play throughout your life.

If we all practiced the *Silver Rule,* originally credited to Confucius, in our interactions with others, we probably wouldn't need to discuss this issue at all:

> ***What you would not want others to do unto you, do not do unto them.***

Abiding by this adage, we certainly wouldn't put others down, stereotype others, treat others unfairly, resent others, or make others the butts of our jokes, since we would not like to have these things done to us.

REFLECTION

Contrast the Silver Rule above with the more widely quoted Golden Rule (*Do unto others as you would have others do unto you*). How do the two rules differ? Is there a difference between telling people what they should do to others and telling people what they shouldn't do to others? Can you think of some examples where you did something to someone that you wouldn't want another person to do to you?

As a nation we have made significant progress in the area of race relations and multi-culturalism, but not nearly enough. Robert Cottrol of Rutgers Law School gives us an optimistic vision for the future [18]:

Perhaps our most important contribution to the twenty-first century will be to demonstrate that people from different races, cultures, and ethnic backgrounds can live side by side; retain their uniqueness; and, yet, over time form a new common culture. That has been the American story. It is a history that has much to tell the world.

> ## I hope YOU will contribute to making this vision a reality!

6.5 ASSESSMENT OF YOUR STRENGTHS AND AREAS FOR IMPROVEMENT

Once you are committed to personal development, you need to start by assessing your strengths and areas for improvement. We talked about this earlier in the chapter when we introduced the concept of a "personal TQM philosophy," and discussed making value judgments about our actions, thoughts, and feelings based on our goals.

Another basis for assessing your strengths and areas for improvement was presented even earlier – in Chapter 1, where we looked at models for viewing your education.

ASSESSMENT BASED ON ATTRIBUTES MODEL

If you recall, the **Attributes Model** indicates the knowledge, skills, and attitudes that you should gain from your engineering education. A personal assessment based on this model involves evaluating how strong you are with regard to each of the following attributes:

An ability to apply knowledge of mathematics, science, and engineering
An ability to design and conduct experiments, as well as to analyze and interpret data
An ability to design a system, component, or process to meet desired needs
An ability to function on multidisciplinary teams
An ability to identify, formulate, and solve engineering problems
An understanding of professional and ethical responsibilities
An ability to communicate effectively
A broad education necessary to understand the impact of engineering solutions in a global and societal context
A recognition of the need for, and an ability to engage in, life-long learning
A knowledge of contemporary issues
An ability to use the techniques, skills, and modern engineering tools necessary for engineering practice

ASSESSMENT BASED ON EMPLOYMENT MODEL

Similarly, you could do a personal assessment based on the **Employment Model**. This model identifies those factors employers use in evaluating you when you apply for a job:

Communicates effectively in a variety of different ways, methods, and media
Possesses the ability to think both critically and creatively
Shows initiative and demonstrates a willingness to learn
Functions effectively on a team
Possesses the ability to think both individually and cooperatively

| Demonstrates an understanding of engineering, science, and mathematics fundamentals |
| Demonstrates an understanding of information technology, digital competency, and information literacy |
| Maintains a positive self-image and possesses positive self-confidence |

ASSESSMENT BASED ON ASTIN STUDENT INVOLVEMENT MODEL

Or you might conduct a personal assessment based on Alexander Astin's **Student Involvement Model**. This model measures the quality of your education as reflected by your level of student involvement in the following areas:

| Time and energy devoted to studying |
| Time spent on campus |
| Participation in student organizations |
| Interaction with faculty members |
| Interaction with other students |

HOW TO DO A PERSONAL ASSESSMENT

Whatever model you choose, doing a personal assessment simply involves rating yourself (e.g., on a scale of zero to ten; ten being highest) on each item. For example, on a scale of zero to ten, how would you rate your ability to work on teams? How would you rate your ability to identify, formulate, and solve engineering problems? How would you rate your ability to communicate effectively? How would you rate your grade point average? To what extent are you involved in student organizations?

For those items that get a high mark, just keep on doing what you are doing. We have a tendency to seek out areas in which we are strong. As a result our strengths are naturally reinforced and there is no need to structure a personal development plan around them.

PERSONAL DEVELOPMENT PLAN

What you need to work on are those areas that got low marks on your personal assessment – the very areas we all tend to avoid. For example, if you do not write well, you try to find classes that require little writing. If you are shy, you tend to avoid people. Avoidance behavior ensures that you will not improve in the very areas that most need your attention.

Realistically, you can't take on all your areas for improvement at one time. So you need to prioritize them in order of importance and choose several of the most important to work on. For each area chosen, create a *personal development plan*. What are you going to do in the next week? In the next month? In the next year?

As an example, if you are shy and lack good interpersonal communication skills, your plan might include some or all of the following actions:

(1) Talk more with people

(2) Discuss your problem with a professional in the counseling center

(3) Take an interpersonal communication course

(4) Read a book on self-esteem

(5) Join the campus Toastmasters Club

(6) Take an acting class

(7) Become involved in a student organization

The time you are in college is the time to work on areas you need to strengthen. You can make mistakes then and the price will be low. If, however, you avoid dealing with those areas in which you need to improve, they will follow you into the engineering work world.

> **There, the price of failure will be much higher!**

6.6 DEVELOPING YOUR COMMUNICATION SKILLS

An important area of your personal development is your ability to communicate. Although you will receive a certain amount of training through your formal education process, it is very likely to fall short of what you really need. In this section, we will explain the necessity for strong communication skills in engineering and suggest ways to supplement your required coursework in this area.

THE IMPORTANCE OF COMMUNICATION SKILLS IN ENGINEERING

Consider the following scenarios:

- As a result of your senior engineering design project, you invent a new Internet search engine. You want to patent, produce, and market the program before someone else beats you to it.

- With engineering diploma in hand, you are ready to launch your professional career, and you have narrowed your job search to four local engineering firms.

- As an engineer who designs and builds vehicles using alternate energy sources, you are assigned to a team to study the durability of a new composite graphite cloth for the vehicle's body. After six weeks of testing the material, your team is ready to make its recommendations to management.

- Shortly after you graduate with your degree in electrical engineering, you are hired into a position at a prestigious research lab, where your primary responsibility is to develop processes for manufacturing integrated circuits used in communications satellites. Four months into your work, you are selected to present your research findings at an upcoming meeting at NASA.

- As a mechanical engineer, you work for a popular amusement park, where you oversee the planning, design, and construction of new rides. In your current project, you supervise a team of 30 engineers and construction workers. Every two weeks, you must submit a written report to your superiors detailing the progress of the project.

At stake in each of these scenarios is not so much your technical expertise as your ability to communicate – both orally and in writing. To patent, produce, and market the Internet device would involve extensive interactions with all kinds of people, groups, and organizations – from the U.S. Patent Office to prospective clients. To land a job after graduation will depend largely on how well you comport yourself and communicate with others (from your resume and cover letters to follow-up phone calls, interviews, and possibly short presentations). As a member of the engineering team at the alternate-energy vehicle company, you will take part in either writing a technical report or presenting your findings to management. The same applies to your meeting at NASA or your progress reports to superiors at the amusement park.

I could easily add many more scenarios to the ones above or create a lengthy list of the writing and speaking demands that you will typically encounter as an engineer. Here are a few:

- Letters, memoranda, and e-mail correspondences
- Design specifications
- Requests for proposals ("RFPs")
- Proposals submitted in response to RFPs
- Contracts, patents, and other government documents
- Progress reports (written and oral)
- Daily work logs
- Instruction manuals
- Technical reports
- Formal presentations (often to non-engineering audiences)
- Project and committee meetings
- Short courses and training seminars
- Guest lectures at engineering schools or professional engineering society conferences
- Publications of findings in professional engineering journals
- Written and oral performance evaluations of subordinates

This list goes on, but I think you see how expansive a role communication skills play in an engineer's world. As a practicing engineer, you will consistently – almost endlessly – draw on your abilities to write and speak effectively.

THE ENGINEERING "DISCOURSE"

The amount and variety of communications, as illustrated above, should be sufficient to convince you of the importance of these skills in engineering. But there is another factor you need to understand in order to appreciate fully the value of strong writing and speaking skills.

In **all** communication transactions, regardless of their importance, length, audience, or medium, your performance will **always** and **inevitably** reflect back on you. Fair or not, your fellow professionals will judge how skillfully you execute every writing or speaking task. If your performance somehow doesn't "measure up," your potential success as an engineering professional will be undermined – and possibly severely limited.

You see, every profession, engineering included, expects certain knowledge, competencies, and an overall presentation of self in order to be accepted into the profession.

James Paul Gee, professor of literacy studies at Arizona State University, explains these professional expectations in terms of a "Discourse" [19]. Gee defines a "Discourse" as

a kind of "identity kit" that includes specific "knowledge, costumes, and expectations on how to act, talk, and write so as to [exhibit] a particular role that others in the Discourse will recognize."

The catch is that, to become a member of a professional Discourse, like engineering, you must be able to demonstrate mastery of **all** of the requirements: "Someone cannot participate in a Discourse unless he or she meets all expectations required in that Discourse. Because Discourses are connected with displays of an identity, failing to display **fully** an identity is tantamount to announcing that you don't really have that identity." In short, "you are either in it or you're not." As we have already seen, communication skills play a major role in the engineering profession. You can be sure, then, that your mastery of these skills will be a pivotal factor in determining your eligibility for membership in the engineering Discourse.

EMPLOYERS WANT MORE

Although the term "Discourse" is unfamiliar to engineering employers and educators, they would certainly understand and agree with the concept behind it. For years, employers have complained about the weak communication skills of the engineering graduates they interview, especially when they find a "star" candidate who qualifies in every way except communications. Not surprisingly, a national survey of over 1,000 engineering employers conducted by the National Society of Professional Engineers revealed that industry's number one concern was to give engineering students "more instruction in written and oral communications" [20]. This conclusion was borne out by a recent excellent study [21] by the Corporate Member Council of the American Society for Engineering Education (ASEE) that indicated that the most important attribute of an early-career engineering professional was:

Communicates effectively in a variety of different ways, methods, and media

Educators, too, have long debated ways to improve their engineering students' communication skills. Pressured both by industry's demands and by ABET's criterion that engineering graduates have "an ability to communicate effectively" (see Chapter 1, Section 1.4), engineering educators everywhere are actively engaged in finding solutions to the problem.

Some engineering schools are developing upper-division, engineering-specific communication courses for their students. To compliment the required composition and speech courses, many institutions are creating writing-intensive courses across the curriculum, and "writing proficiency exit exams." This increased focus on students' communication skills will hopefully produce better prepared graduates. In most cases, though, the engineering curriculum is already so full that additional requirements are not possible. So the debate continues.

THE ENGINEERING STUDENT AS COMMUNICATOR: A PROFILE

Take a moment to reflect on your own communication skills. How would you rate your writing? How do you feel about speaking before audiences or giving oral presentations? Chances are that, as an engineering major, you view writing and public speaking with little enthusiasm. They are probably your least favorite and perhaps weakest subjects, and the fewer courses you must take in them, the happier you will be. After all, you likely chose engineering because your strengths lie in math and science; your engineering curriculum centers largely on developing these strengths; and a "good" engineer is one who excels in these areas.

Although you now understand the need for strong communication skills, you may lack confidence or feel discouraged about your ability to improve them. Many new engineering students do, especially those whose native language is not English. To make matters worse, you are bound to encounter fellow engineering students whose communication skills clearly outshine your own. How, you wonder, can you ever learn to communicate as well as they do? Your required coursework in communications will help, but will that be enough? Probably not.

The answer is that you *can* become an effective communicator. The skills are not all that difficult to master, but they do require two commitments from you:

(1)   A positive attitude
(2)   A 4-5 year plan to ensure regular practice

DEVELOPING A POSITIVE ATTITUDE

Let's address attitude first. The profile above, highlighting the fear, intimidation, discouragement, and perceived inability that many engineering students feel about their communication skills, is quite understandable. Most people, regardless of profession, dislike writing of any kind and abhor public speaking. It's not surprising. Writing and speaking are like "baring your soul" to others – often strangers. To some extent, you can hide weak writing skills by hiring an editor or ghostwriter. You also have the opportunity to proofread and revise – again and again – before making a written document public. Speaking before an audience, however, offers no such safety nets. Like a ballet dancer or tight-rope walker, you need only one wrong step to ruin your performance.

While understandable, the fears associated with writing and public speaking are counterproductive – particularly for you, as future engineers. I can't tell you how many times I have seen promising engineering students try to capitalize on their technical knowledge as a way to compensate for inadequate communication skills. Because they feared or disliked writing and speaking, they did as little as possible in college to develop these skills. When, as seniors, they began their job search, they repeatedly lost out to others who had wisely balanced their engineering education with training and practice in communication.

It would be great – wouldn't it? – if we could change our negative attitudes about writing and speaking instantaneously. Just one "ZAP!" from Batman or the wave of a magician's wand and – voila! – we can't wait to write a technical report or make a presentation before a group of executives. But, while we may lack such magical swiftness, we do have the capacity to change negative attitudes into positive ones. In terms of writing and speaking, you need just one little success – an *"A"* grade on a composition or a compliment from a professor about your contributions in class – to start the process. Make one of these small challenges a goal of yours this semester, and watch how your attitude about writing and speaking begins to change.

PLANNING TO IMPROVE YOUR COMMUNICATION SKILLS

You've heard the expression that "practice makes perfect." It's true. No matter how adept or advanced you may be in a skill, you need to practice it regularly to maintain it. Just ask any professional athlete.

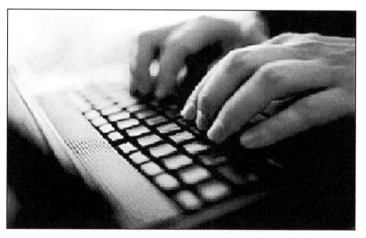

Developing strong communication skills takes <u>years</u> (not days, weeks, or even months). That's why you need to start TODAY. The longer you delay tending to them, the weaker they will be when you graduate – and the longer it will take you to prove your eligibility to join the engineering Discourse.

What should you do? Lay out a 4-5 year plan that includes as many courses that focus on communication skills as you can work into your program. Build in your required courses first, like freshman composition and speech communication. Then browse through your university catalog for additional communications courses: advanced writing and speech courses, business communication courses, psychology/human relations courses, theater courses, "writing-intensive" courses, and so forth. See which of these courses could be used to meet your general education requirements. You may also have some free electives that you can devote to developing your communication skills. Another option would be to take a few extra courses in this important developmental area that don't meet graduation requirements. An ideal goal would be to take one course in this area each term during your four-year program.

Once your plan is complete, put it into action. Approach each class with the same enthusiasm and interest you bring to your engineering courses. Sit in the front of the classroom. Be inquisitive: Ask questions. When faced with a writing assignment, do it early so that you can get feedback from your instructor or tutor, and rewrite it based on their suggestions. Do the same for oral assignments. Practice an upcoming presentation before a group of friends or classmates. Even better, video record yourself and then evaluate your performance. We're always our own toughest critics!

Beyond formal coursework, look for extracurricular opportunities to write and speak. Practice your writing by keeping a journal or corresponding with friends via email. Write a poem or short story. *Write a critique of this book and send it to me!* Above all, **read** – anything and everything. Research in language development has shown that reading is a significant way to improve writing skills. Read the newspaper, magazines, technical journals, and novels. Set goals for your reading, particularly during breaks between school terms. Take the time you would normally watch television or play video games and use it for recreational reading.

Develop your oral communication skills with the same vitality and engagement, both in and out of the classroom. A speech class will introduce you to the field of rhetoric and give you practice in the rhetorical modes of discourse (e.g., narrating, describing, analyzing, persuading, and arguing). Psychology courses will teach you the principles of human relations, group dynamics, and cross-cultural communications. A theater course will give you instruction and practice in effective delivery.

Extracurricular activities are also plentiful and beneficial. Getting involved in engineering student organizations or school athletic programs will go far in building your interpersonal and teamwork skills. Running for student offices and holding positions in the student body government will strengthen your public speaking and presentation skills. Outside of school, scrutinize speakers you hear on television and radio. Study the techniques of famous speakers, like Dr. Martin Luther King, Jr. and John F. Kennedy, or *any* individual whose speaking skills you admire. Take stock of their strengths and weaknesses, and try to incorporate some of their strategies into your own speaking repertoire.

CONCLUSION

Whatever avenues appeal most to you to develop your communication skills, what's most important is that you **DO SOMETHING** and **START TODAY**. A 4-5 year plan begun *now* will help keep you on track. If you sincerely commit to this plan, I guarantee that your most defeatist attitude will change for the better. Watch your self-confidence and poise grow. Note how your fellow engineering students start to admire you and seek your "secret" to success. Don't be surprised when prospective employers start vying for your attention.

6.7 LEADERSHIP AND TEAMWORK

A *team* is defined as two or more people who interact regularly and coordinate their work to accomplish a mutual objective.

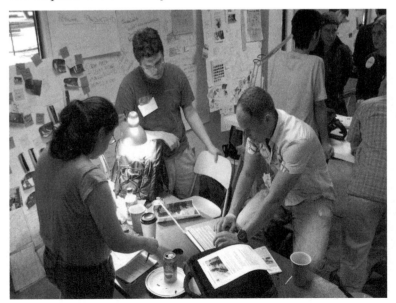

Nothing of significance is ever achieved by an individual acting alone. Think of impressive human achievements: climbing of Mt. Everest, launching a fighter jet from an aircraft carrier, placing a human on the moon, building the Golden Gate Bridge. We're all aware that these achievements required significant teamwork. But even things that appear to have been done by individuals – Jonas Salk developing the polio vaccine, Daniel Boone blazing the wilderness trail, Charles Lindbergh flying solo across the Atlantic, Bill Gates creating Microsoft, Mohammed Ali winning the world heavyweight boxing championship, Mother Teresa ministering to the poor – were really team efforts. Each of these individuals required others to achieve their results.

If you, as many students do, prefer to work alone, now is the time to get over it. Virtually all engineering work is teamwork. You will be involved on a team – as a team member or leader – throughout your engineering education, your career as an engineering professional, and your life.

One indication of the importance of teamwork is the fact that, as you learned in Chapter 1, one of the 11 attributes ABET requires of engineering graduates is:

an ability to function on multi-disciplinary teams

Unless you have demonstrated an ability to work on teams, you aren't even eligible to be awarded a degree in engineering. I therefore encourage you to take every opportunity to develop your effectiveness as both a team member and a team leader.

As a student, you will be <u>required</u> as part of your formal coursework to participate extensively on two types of teams:

- Laboratory groups
- Engineering design project teams

But you will have the opportunity to <u>elect</u> to participate on other types of teams including:

- Study groups
- Engineering student design competitions (e.g., steel bridge truss competition, human-powered vehicle, etc.)
- Research teams
- Service project teams
- Student organizations and student government

Furthermore, as you go through your life, you will undoubtedly establish a broader context for "teams" you are involved in, including your family members, neighbors, colleagues at work, community volunteers, or anyone you decide to work with toward a common purpose.

Comprehensive coverage of the subject of teamwork and leadership is beyond the scope of this book. In this section, we give an overview of the subject, with the goal of persuading you to learn and practice teamwork and leadership skills.

Many complete books are available on the subjects [22, 23, 24], as are courses you will find on your campus (in the business school or the communication studies department). The following sections provide an overview of the subject of teamwork and leadership.

PRINCIPLES OF TEAMWORK

Following are some general principles that govern every high- performing team.

Purpose – Teams need a clear, well-defined vision of what needs to be accomplished.

Synergy – By combining efforts and talents, teams can outperform any individual.

Cooperation – Team members must be willing to subordinate their self-interests on behalf of the team's purpose.

Roles – Each team member contributes optimally when in the correct, clearly-assigned role.

Difficulty – The more challenging the goal, the more important the teamwork is.

Motivation – Team members need to find ways to spark the team's greatest possible accomplishments.

Weakest Link – The team's results will be limited by the performance of its weakest member.

Attitude – Poor attitudes can spoil the greatest of talents.

Trust/Reliance – Team members need to be able to rely on one another.

Discipline – Team members must be prepared to make the necessary sacrifices to do what needs to be done.

Focus – Teams need a system to track their progress and keep them focused on their goal(s).

Values – Team members should have shared values about their project.

Leadership – The team needs to have effective leadership.

Morale – Team members should feel good about being a member of their team.

Planning and Resources – The team should effectively plan and use resources.

Decision-Making – The team needs to be able to make good decisions.

REFLECTION

Have you ever been a member of a team of any kind? If so, did the team exhibit the attributes listed above? Do you feel as though you could help bring some or all of these attributes to any teams you are part of in the future? In what ways? Which attributes do you think are most important? Could you be part of a high-performing team that lacked one or more of these attributes? If so, explain?

ATTRIBUTES OF AN EFFECTIVE TEAM LEADER

As every ship needs a captain, every team needs a team leader. Depending on the context, the team leader might be appointed (e.g., by your professor) or elected by team members. The purpose of the leader is to direct the actions of the team to achieve its goal.

Some of the important attributes of an effective team leader include:

- Willingness to lead and take charge
- Ability to keep the team focused on its purpose
- Ability to set goals, priorities, and standards of performance
- Proficiency at being a team builder
- Ability to plan appropriately/accordingly
- Able to run productive meetings
- Ability to communicate effectively
- Ability to promote harmony and inspire trust
- Ability to foster high levels of performance by team members

No matter what their traits or skills, leaders carry out their roles in a wide variety of styles. Three of the most common are:

1) Autocratic – Leader makes decisions independently with little input from team members.

2) Democratic – Leader offers guidance but also encourages strong participation from team members.

3) Laissez-Faire – Leaders offer little guidance and leave decision-making up to team members.

REFLECTION

Review the attributes of an effective team leader. Do you think you would be an effective team leader? Which of the attributes of a team leader would you be strongest in? Which would you be weakest in? What could you do to develop those attributes in which you need improvement?

In sum, the leadership style portrayed in the cartoon below is <u>NOT</u> something you want to aim for!

© Randy Glasbergen / glasbergen.com

"If we want to succeed as a team, we need to put aside our own selfish, individual interests and start doing things my way."

CHARACTERISTICS OF AN EFFECTIVE TEAM MEMBER

Some of the characteristics of an effective team member are:

- Supports and helps the team leader succeed
- Understands and supports the team mission, purpose, and goal
- Subordinates self-interest on behalf of the team's purpose
- Welcomes being a member of the team and works to get to know and build trust with other team members
- Communicates openly and honestly
- Respects differences and diversity in team members
- Works to elicit the ideas of others; listens to understand others' points of view

- Views conflict as useful and necessary; works toward consensus
- Is reliable; follows through on tasks; meets deadlines
- Is willing to work hard, often "beyond the call of duty," for the success of the team

STAGES OF TEAM DEVELOPMENT

You can't expect a newly formed team to perform at a high level from very start. Team formation takes time and effort, and team development usually follows distinct stages as the team moves from being a group of strangers to becoming a united team with a common goal.

The Tuckman Forming Storming Norming Performing Adjourning Model [25] is a highly regarded and widely accepted model of the stages that are inevitable for a team to grow to the point where they are functioning effectively together and delivering high quality results.

Once a team leader is appointed or elected, the following five stages in the team's development generally take place.

STAGE 1: FORMING. The team may be formed in an initial meeting where team members meet each other and share information about their backgrounds, interests, and experience. At this point the role of the team leader is vital, as he or she will provide information about the project's objectives and goals, lead discussion about the scope of the task, and solicit ideas about how the team will work. This stage focuses mainly on routine issues such as team organization, assignments, and scheduling.

STAGE 2: STORMING. Once the work begins, the team moves into the "storming" stage. In this stage team members typically compete with each other for status and acceptance of their ideas. They undoubtedly will have differences about what should be done—an area of potential conflict within the team. Guided by an effective team leader, members learn how to solve problems together and settle into their roles and responsibilities. Team members must learn to listen to each other and respect one another's differences and ideas. At this state, members should have become team-focused.

STAGE 3: NORMING. During this stage team members begin to work together as a team. Roles and responsibilities are clear and accepted and decisions are "group approved." Team members have learned to respect each other's opinions, and they begin to trust each one another's input and assistance. Rather than competing against each other, they now help each other in pursuit of their common goal.

At this stage the team starts to make significant progress on the project. Since team members are collaborating, the team leader begins to function more like a coach.

STAGE 4: PERFORMING. In this stage the team should be functioning at a very high level. Group identity, loyalty, and morale should be strong, with the team highly motivated to get the job done. Team members should trust and rely on each other, making decisions and solving problems quickly and effectively. Team members do not need the oversight from the team leader that was necessary earlier.

STAGE 5: ADJOURNING. Adjourning takes place once the task is completed. If the project was successful, lessons learned should be documented and, of course, the team's success celebrated.

As team members move on to new endeavors, they will be enriched by the successes and challenges/obstacles that will help them in future projects.

6.8 MENTAL AND PHYSICAL WELLNESS

To be productive and happy, it is important that you take care of your mental and physical well-being. With the rigors and demands of being a student – and, later an engineering professional – it is easy to ignore these areas of your life. But that is a big mistake. Tending to these personal needs is a must.

Many people are not aware of the connection between one's physical and emotional health. The truth is they are strongly interrelated: Our physical well-being greatly affects our emotional state – and vice versa. For example, one of the best remedies for emotional stress is vigorous physical exercise. And I'm sure you've noticed that when you are mentally "up," you tend to feel good physically, whereas if you're emotionally down, you often feel physically fatigued or even get sick.

TIPS FOR GOOD HEALTH

Since each of us is so unique and our emotional and physical states so complicated, this section is only meant to offer you a few ideas. Most obviously and most importantly, to expect a high level of mental and physical health, you must:

  ➢ **Eat nutritionally**
  ➢ **Engage in regular aerobic exercise**
  ➢ **Get adequate sleep**
  ➢ **Avoid drugs**

EAT NUTRITIONALLY. What you eat significantly affects your physical and mental state. A proper diet consists of fresh fruits and vegetables, lean meat and fish in moderation, and whole-grain products. Avoid processed foods, fatty foods, and sugar. Not only will you feel better now but you'll reduce your chances of heart attack, cancer, and other diseases later.

ENGAGE IN REGULAR AEROBIC EXERCISE. Regular vigorous exercise, in which you get your heart rate above 130 beats per minute for more than 20 minutes at least three times a week, is essential to good physical condition. If you're not already engaged in some form of exercise, you should consider taking up jogging, brisk walking, swimming, biking, rowing, dancing, spinning, or any activity you commit to doing regularly.

GET ADEQUATE SLEEP. Different people require different amounts of sleep, and the amount needed will likely change as you grow older. Only you can determine how much sleep you need. Just remember that your work efficiency will decrease if you are getting either less or more sleep than you need.

AVOID DRUGS. Drugs are abundant in our society. Some, such as caffeine, alcohol, and nicotine, are legal; others, such as marijuana and cocaine, are illegal. Regardless of their legality, all can be harmful and my advice to you is simple: Avoid them. Not only do drugs detract from your physical and mental health, they also can greatly interfere with your ability to study.

BALANCING WORK AND PLAY

To ensure a healthy mental state, you need to **strike a balance between immediate and future gratification**. By seeking too much immediate gratification (and therefore not getting your work done), you are likely to feel guilty. You'll probably then worry about the fact that you are not studying, putting yourself into a mental state in which you cannot study.

On the other hand if you work too much, too long, or too hard, you begin to feel deprived. Feelings of deprivation and resentment care bound to sabotage your commitment to your studies. You may begin to doubt whether the sacrifice is worth it.

What you need to establish is a healthy balance between work and play. One approach is to tie work and play together through a system of rewards. Rewards can be small things, like taking a break, going for a walk, watching your favorite TV show, taking an hour for recreational reading. They could also be larger things, such as going to a party, buying new clothes, or planning a weekend getaway.

> *The point is that, rather than take the view that the work you are doing will not have a payoff until far into the future, you provide yourself with frequent and immediate rewards for your hard work.*

MANAGING STRESS

Interestingly, the term *stress* was borrowed from engineering by Dr. Hans Selye, an early pioneer in the area of stress management. Selye defined stress as "the response of the body to any demand made upon it to adapt, whether that demand produces pleasure or pain." [26]

Stress can be imposed externally or internally. Certainly, the announcement by three of your professors that each has scheduled a midterm exam on the same day can create external stress. Causes of internally imposed stress include unmet expectations, high performance standards, and unrealistic demands you place on yourself.

Stress can be either positive or negative. *Eustress* is a positive form that motivates individuals to attain high levels of performance. For example, the "butterflies" a football player experiences before a big game can produce inspired play. *Distress* is the negative form of stress. It distracts you from being the best you can be. It can debilitate both your physical and mental health.

Some common causes of distress include worry, frustration, anxiety, and depression. Frustration is our response to being prevented from gratifying certain impulses or desires. For example, it would be frustrating if you were unable to enroll in a mathematics course you needed to take. Worry and anxiety are closely related. Both are your response to a perceived threat. Anxiety is a somewhat stronger emotion. You become worried if your roommate fails to come home for several days. You become anxious when you develop pains in your chest. Depression is an extreme form of worry and is an emotional condition characterized by feelings of hopelessness and inadequacy.

Each of these emotions is a potential stressor. Of course, stressors do not affect everyone the same way. What would cause stress for one person may not even bother another.

What is clear, though, is that stressors are directly related to your self-esteem. If you feel competent and worthy, you are in a good position to handle stress. Your reaction to stressors is also tied to your sense of control over your life. For individuals who lack self-esteem and do not feel in control of their lives, stress can wreak all kinds of emotional and physical havoc.

Whether *eustress* or *distress*, Selye demonstrated that stress stimulated our "fight or flight" response. This is an instinctive physical reaction to threat, either physical or psychological, which we inherited from our ancient ancestors. Under stress, more blood is diverted to the brain and muscles for clearer thinking and quicker reflexes, the heart rate accelerates, blood pressure rises, and respiration rate increases. For engineering students, this response is not appropriate, since you will need to neither "fight" nor "take flight." The stressors we're talking about are mental or emotional ones, so the concepts behind "fight or flight" are, at best, metaphoric. Still, it is interesting to understand the interactions between perceived stressors and one's response to them.

To bolster your effectiveness as an engineering student, it is important that you learn to cope with and manage stress. HelpGuide.org (www.helpguide.org/topics/stress.htm), a non-profit mental health organization, puts forth 14 healthy ways to reduce stress and relax and recharge:

(1)	Go for a walk.
(2)	Spend time in nature.
(3)	Call a good friend.
(4)	Sweat out tension with a good workout.
(5)	Write in your journal.
(6)	Take a long bath.
(7)	Light scented candles.
(8)	Savor a warm cup of coffee or tea.
(9)	Play with a pet.
(10)	Work in your garden.
(11)	Get a massage.
(12)	Curl up with a good book.
(13)	Listen to music.
(14)	Watch a comedy.

It is important that you use practical strategies such as these to prevent the burn-out that can come from unchecked or unresolved stress. In the long term, proper nutrition, regular exercise, relaxation, and good time management will help to keep your stress level low.

If all else fails, it is important that you seek counseling or medical treatment. Extreme stress can lead to severe physical incapacitation – often to the point of prolonged diseases or even death.

6.9 MOTIVATING YOURSELF

We will conclude this chapter on student development with several inspirational messages. If you accept the premise of behavior modification that you hold the reins for becoming the best you can be, the messages of others who have succeeded will reinforce your motivation to do the same.

Another Personal Story

Sometimes I wish I could bottle and sell the feeling I have about the value of my education. I owe almost everything of quality in my life to my education. I have had so many unique and rewarding experiences, so many challenges, so many opportunities. I've been paid well to do work I enjoy. I have been able to travel, write, speak, teach, and influence others. I have gotten to know many interesting people. I have had options and choices and control over where I live and what I do. I can hardly imagine going through life without an education.

Often I wonder why I have been so fortunate. How is it that I went so far with my education? Where did my motivation come from? Actually, it is as clear as day. My mother gave it to me. And she did it in such a dramatic way that, as I look back, there was no possibility I would not go to college and no possibility I would not succeed. If I had failed, I would have let down my mother's number one dream.

How did she motivate me so strongly? When I was two years old, my father died. He had just changed jobs, so my mother received only a very modest pension of $15 a month

As a single parent, my mother could have really used that money, but every month she put $3.75 with that $15 and with the $18.75 she purchased a $25 United States Savings Bond for my education. Every month she took me to the bank to put the savings bond in her safe deposit box and told me that the money was for my college education.

Your parents or guardians may have sent you similar messages about the importance of education. If they did, you are indeed fortunate. You are very likely to succeed.

Not all students have parents or guardians who sent them positive messages about the value of a college education. I have known many students, in fact, whose parents were opposed to their going to college, preferring instead that they work to help support the family.

However, regardless of any messages you received growing up about the value of a college education (and the consequent success it will bring you), you are now an adult. You can think for yourself. You can define your own reasons for wanting to get your education. You can motivate yourself.

We already addressed the idea of motivating yourself through an awareness of the rewards and opportunities an engineering career offers. In Chapter 2, we discussed my "top ten" list of what it will mean to the quality of your life if you are successful in completing your engineering degree. The following sections present additional perspectives regarding the value of your education.

"NO DEPOSIT, NO RETURN"

When I was Director of the Minority Engineering Program at California State University, Northridge, we wore a motivational button with the logo shown to the right.

The clenched fist holding the slide rule (a symbol of engineering in years past) represents the power that comes to an individual through a technical education. The motto "No Deposit, No Return" reminds us, however, that this power does not come without a "deposit." Getting a college education is not easy, and majoring in engineering is even more challenging. Without a doubt, the "deposit" you must make as an engineering major is a rigorous, demanding one. But I guarantee that you will be paid back for the investment many, many times over.

JESSE JACKSON'S "EXCEL" MESSAGE

The Reverend Jesse Jackson has spent much of his career motivating young students. His *Push for Excellence* program had a significant impact on African-American youth in Chicago and across the nation.

Reverend Jackson's basic message is that you should strive to excel in everything you do. But what does it mean to excel? Very simply, it means:

Do your personal best!

If your "personal best" produces a "*C*" grade in a course, then you have excelled in that course. But how many of us do our best? Do you?

There are reasons why you should strive to excel – to do your best. We have already discussed some:

➤ It will enhance your self-esteem.

➤ It will be good for society.

➤ You will have a challenging and rewarding career.

Rev. Jesse Jackson Can you think of others?

INSPIRATIONAL AND MOTIVATIONAL QUOTATIONS

I like inspirational and motivational quotations. You have probably gathered that by the fact that each chapter in this book starts with a quotation and others are dispersed throughout the chapters. Many of these quotations speak to me personally, reminding me of what's important. And they offer so much wisdom. One from former British Prime Minister Benjamin Disraeli aptly notes that:

Much of the wisdom of the wise and the experience
of the ages are perpetuated by quotations.

You can find a tremendous resource of more than 500 of the most famous quotes at:

www.discovery-press.com/discovery-press/studyengr/quotes.htm.

There are so many wonderful quotes at that website that it's hard for me to choose favorites. Examples of ones I like are:

"Do not go where the path may lead; go instead where there is no path and leave a trail."
– Ralph Waldo Emerson

"You've got to get up every morning with determination, if you're going to go to bed with satisfaction." – George Lorimer

"When one door of happiness closes, another opens, but often we look so long at the closed door that we do not see the one that has been opened up for us." – Helen Keller

"The tragedy of life doesn't lie in not reaching your goal. The tragedy lies in having no goals to reach." – Benjamin Mays

"Whether you think you can or think you can't, you're right." – Henry Ford

"Do first things first, and second things not at all." – Peter Drucker

"Tell me and I forget. Teach me and I remember. Involve me and I learn." – Benjamin Franklin

Reaping the potential benefits of quotations will take time and effort on your part. I encourage you to read as many of the 500 ones at the website above as possible. But reading a quotation is not enough. Select a few that particularly speak to you. Reflect on each one's meaning and the changes in your attitudes or behaviors each calls on you to make. Write each quotation on an index card and post the cards where you can readily see them. Reread the quotations often for seven days. Memorize several and use them in conversation with others. Above all, be guided by Benjamin Franklin's above astute connection between involvement and learning.

REFLECTION

Reflect on Peter Drucker's quotation, "Do first things first, and second things not at all." What does this mean to you? Do you see something of yourself in this quotation? Are you a person who prioritizes the tasks in front of you and does the top priority item first? Or do you choose to do things that are easy or you enjoy? Once you have completed your top priority task, do you go to your second priority task? Or do you reprioritize the remaining tasks?

POWER OF POSITIVE THINKING

I like to use an analogy between jogging and going to college. People take up jogging because they perceive certain benefits. They expect to live longer, feel better, breathe more easily, and lose weight. Initially they may dislike the experience of jogging, suffering through it solely for the end result. Eventually, however, most joggers learn to enjoy the experience. They come to enjoy the physical elements – the rhythmic cadences of moving and breathing, the harmony between body and mind – and they find that long periods of jogging can lead to a particularly unique experiences: the so-called "runner's high," a heightened sensitivity to the world around them, along with an ability to think more creatively.

As an engineering student, you can liken yourself to a jogger. At first you may resent or dislike the college experience but you persevere because of the future benefits you anticipate: career opportunities, money, social status, security. But you also will come to appreciate your time in school, not only for the benefits it promises, but for the experience itself.

If, like the novice jogger, you find that you dislike school, you are not focusing on the positive aspects of being a college student. You need to recognize that you have created an attitude that may have nothing to do with reality. In fact, you probably are in the best situation of your life and just not aware of it. Surely you have heard people say that their college years were the best years of their lives. Why do you suppose they say this?

If you do have a negative attitude toward school, now is the time to change it. For it's more than likely that you are neither performing at your peak effectiveness nor enjoying what should be the most exciting, rewarding time in your life. Learn to focus on the positive aspects of being a college student, like the ones listed below.

GROWTH PERIOD. As a college student, you are in a significant growth period. One indication of this is the way in which you are outgrowing your friends from high school who are not going to college. Probably never in your life will you be in such an intense period of learning and personal growth as when you are in college.

EXPOSURE TO PEOPLE. College puts you in an extremely people-oriented environment. Never again will you be with so many people of the same age and interests. You'll find that many of the friends you make during your college years will be important and helpful to you throughout your life.

MANAGER OF YOUR TIME. As a college student you are working for yourself. You have no boss, no one to tell you what to do. Except for your class time, you are pretty much free to manage your time and affairs.

STARTS AND STOPS. School starts and stops, somewhat like the running of a race. When the race starts you put out a great deal of effort, maybe more than you would like to, but you do so because you know it will end. When it does end, you then have an extended period of time for rest and rejuvenation – a break you will not have once you start working as an engineer.

When you learn to appreciate these and other unique aspects of being a college student, you will see an improvement in your academic performance. Remember that:

> *Positive attitudes bring positive results! Negative attitudes bring negative results!*

REFLECTION

Consider the following quote from Winston Churchill: "The pessimist sees difficulty in every opportunity. The optimist sees the opportunity in every difficulty." Which statement best describes you? Do you tend to see difficulties in every opportunity? Or do you tend to see opportunities in every difficulty?

SUMMARY

This chapter focused on your personal growth and your development as a student. Because change is such a critical factor in personal growth and student development, we first discussed the psychology of change. We particularly noted how people's receptiveness to change has grown dramatically in recent years – a credit owed largely to U.S. business and industry's embracing the philosophy of "continuous improvement" and its derivative program, Total Quality Management (TQM).

We then proposed a personal adaptation of TQM, which we called "student development." The particulars of each individual student's development plan will certainly vary, but all are based on the following premises:

(1) Your goals set the context and direction of your personal development plan.

(2) You monitor the progress of your personal development by a measurement system that analyzes your actions, thoughts, and feelings.

(3) You not only implement the plan and track your progress; you also continuously revise the plan to ensure your "continuous improvement."

Next, we discussed mechanisms for human behavioral change and presented one based on behavior modification theory: a three-step process that is the most practical and accessible for students. Those three steps entail (1) knowledge, (2) commitment, and (3) implementation. As part of this presentation, we also discussed barriers to change that you are likely to encounter as you move through the three steps.

A necessary backdrop to behavioral and attitudinal changes is self-understanding, so we discussed various models of human behavior to help you understand yourself better.

Maslow's Hierarchy of Needs showed you what basic human needs must be met before you can undertake a plan of self-improvement. Along with this discussion of basic needs, we pointed out the role self-esteem plays in the process of personal growth and change.

The Myers-Briggs Type Indicator (and the closely related Keirsey Temperament Sorter II) was presented as another way to understand yourself. These instruments give key insight into personality types.

The topic of understanding others and respecting differences in people was also discussed. An important component of your personal growth is your effectiveness in working with people who differ from you.

Next we discussed assessment as a tool to identify and work on areas in which you need improvement. We focused on three particular areas that are often overlooked in personal development: communication skills, teamwork and leadership, and physical and mental wellness. In this latter area, we emphasized the importance of managing stress.

We concluded by offering multiple motivational messages and quotations to support your commitment to the difficult process of behavioral and attitudinal change. In conjunction with these messages, we highlighted the positive aspects of being a college student.

The intention of this chapter was to stress how change leads to personal growth, how personal growth leads to the accomplishment of goals, and how the accomplishment of goals leads to success in school, in the engineering profession, and in life in general.

REFERENCES

1. Deming, W. Edward, *Out of the Crisis*, The MIT Press, Cambridge, MA, 2000.

2. Chopra, Deepak, *The Seven Spiritual Laws of Success*, Amber-Allen Publishing & New World Library, San Rafael, CA, 1994.

3. Besterfield-Sacre, M.E., Atman, C.J., and Shuman, L.J., "Student Attitudes Assessment," *Journal of Engineering Education*, 87(2), April 1998, 133-141.

4. Gray, Albert E.N.*, The Common Denominator of Success*, Tremendous Life Books, 2008. (www.discovery-press.com/discovery-press/studyengr/CommonDenominator.pdf)

5. Lehman, K., *The Birth Order Book: Why You Are the Way You Are*, Revell Books, Grand Rapids, MI, 2004.

6. Maslow, Abraham, *Motivation and Personality*, Harper and Row, New York, NY, 1970.

7. "Toward a State of Esteem: The Final Report of the California Task Force to Promote Self-Esteem and Personal and Social Responsibility," California State Department of Education, Sacramento, CA, 1990.

8. Branden, Nathaniel, *The Six Pillars of Self-Esteem*, Bantam Books, New York, NY, 1994.

9. Myers, David G., *The Pursuit of Happiness: who is happy – and why*, W. Morrow, New York, NY, 1992.

10. Jung, C.G., Adler, G., and Hull, R.F.C., *Psychological Types (Collected Works of C.G. Jung Volume 6)*, Princeton University Press, Princeton, NJ, 1976 (originally published in 1921).

11. Briggs, K.C. and Myers I.B., "Myers Briggs Type Indicator, Form G," Consulting Psychologists Press, Palo Alto, CA, 1977.

12. Keirsey, David, *Please Understand Me II: Temperament, Character, Intelligence*, Prometheus Nemesis Book Company, 1998.

13. Tieger, Paul D. and Barron-Tieger, Barbara, *Do What You Are: Discover the Perfect Career for You Through the Secrets of Personality Type*, Litte, Brown and Company, 2007.

14. Wankat, P.C. and Oreovicz, F.S., *Teaching Engineering*, McGraw-Hill, 1993.

15. Herrmann, Ned, *The Creative Brain*, Ned Herrmann/Brain Books, Lake Lure, NC, 1990.

16. *Profiles of Engineering & Engineering Technology Colleges,* American Society for Engineering Education (ASEE), Washington, D.C., 2012.

17. Widnall, Sheila E., "AAAS Presidential Lecture: Voices from the Pipeline," *Science*, Vol. 241, pp. 1740-1745, September, 1988.

18. Cottrol, Robert, "America the Multicultural," *The American Educator,* Winter, 1990.

19. Gee, James Paul, "Meditations on Papers Redefining the Social in Composition Theory," *The Writing Instructor*, Summer 1989.

20. "Report on Surveys of Opinions by Engineering Deans and Employers of Engineering Graduates on the First Professional Degree," Sustaining University Program, National Society of Professional Engineers, Publication No. 3059, Alexandria, VA, November, 1992.

21. Hundley, Stephen P., Brown, Lynn, and Jacobs, Alan, "Attributes of a Global Engineer: Field-Informed Perspectives, Recommendations, and Implications," Presented at 13th Annual Conference of the American Society for Engineering Education, San Antonio, Texas, June, 2012.

22. Maxwell, John C., *The 21 Irrefutable Laws of Leadership: Follow Them and People Will Follow You*, Thomas Nelson, 2007.

23. Drucker, Peter F. *The Practice of Management*, Harper Business, Reissue Edition, 2006.

24. Parker, Glenn M., *Team Players and Teamwork, Completely Updated and Revised: New Strategies for Developing Successful Collaboration*, 2nd Edition, Jossey-Bass, 2008.

25. Tuckman, Bruce, " *Psychological Bulletin* 63 (6), 1965.

26. Selye, H., *Stress without Distress*, J.B. Lippincott, Philadelphia, PA, 1974.

PROBLEMS

1. After doing some research, write a 500- to 750-word essay on "Total Quality Management." How does TQM compare to the management approaches you have observed in the university you attend, organizations you belong to, or companies you have worked for?

2. For the next week, write down any negative thoughts you have. Describe at least one non-productive action that is likely to result from each negative thought. Also write down a positive thought you could substitute for the negative thought. Describe at least one productive action that is likely to result from each positive thought.

3. Scrutinize your behavior in the past week and then write down five non-productive actions you have done. Why did you choose these non-productive actions?

4. When we receive a stimulus (e.g., seeing food), we often act (e.g., eating the food) and then think about our action (e.g., "I shouldn't have eaten the whole thing."). How can you change this order (stimulus ⇒ action ⇒ thought) in a way that would lead you to choose productive actions more frequently?

5. Add ten examples of productive actions (actions that will enhance your academic success) to the list of seven given in Section 6.2.

6. Do you think behavior modification could work for you? Why? Do you think counseling/therapy would benefit you? Why?

7. Convert the following negative thoughts to positive ones by finding a higher context from which to view the situation that led to the negative thought:

    a. I wish I were taller.
    b. I'm homesick.
    c. I don't have any friends.
    d. My chemistry lectures are boring.
    e. I don't know if I like engineering.
    f. I wish I could find a better roommate.
    g. I don't have time to exercise regularly.

8. Go to the following website and print out the survey form there (www.discovery-press.com/discovery-press/studyengr/attitudesurvey.pdf). Complete the "Freshman Engineering Student Attitudes Survey." Review the results and identify any negative attitudes you have. What criteria did you use to make the determination? Pick three important negative attitudes that you hold and try to change them by working through the process described on Pages 170-171. Write a one-page statement on what you learned from this exercise.

9. Consider your view about life-long learning. Are you excited or discouraged about the prospect of learning throughout your career? Prepare a 2-3 minute presentation for your *Introduction to Engineering* course on why life-long learning is particularly important for professional engineers.

10. Consider the following productive behaviors:

    a. Studying collaboratively with other students

b. Devoting significant time and energy to studying

c. Preparing for each lecture

d. Studying from class to class rather than from test to test

e. Making effective use of professors outside of the classroom

f. Practicing good time management principles

g. Immersing yourself in the academic environment of the institution

h. Actively participating in student organizations

Answer the following questions about each behavior:

(1) Do you have adequate knowledge about each behavior?

(2) Have you made a commitment to the behavior? If not, why?

(3) Are you implementing the behavior? If not, what is keeping you from doing so?

11. Read Albert E.N. Gray's presentation "The Common Denominator of Success." (You can find it at: www.discovery-press.com/discovery-press/studyengr/CommonDenominator.pdf). Write a one-page review of the presentation. What do you think of the ideas presented there? Do they stand the test of time? Do you agree with the "common denominator of success" that Mr. Gray identifies?

12. Review Abraham Maslow's Hierarchy of Needs. How well are you meeting your needs at each level? Think up one or two ways to better meet your needs at each level. Do them.

13. Research the term *self-actualization*. What does it mean? How strong is your need for self-actualization?

14. Consider the ten items listed in Section 6.3 that correlate with healthy self-esteem. Write down a brief definition of each item. How many of the items would you use to describe yourself?

15. Consider the ten items listed in Section 6.3 that correlate with poor self-esteem. Write down a brief definition of each item. How many of the items would you use to describe yourself?

16. Based on the results of Problems 14 and 15, would you say you have a healthy or poor self-esteem? Explain. What could you do to improve your self-esteem?

17. Write a paragraph describing yourself in terms of the four personality indicators that are measured through the Myers-Briggs Type Indicator (MBTI). From this analysis, what MBTI indicator (e.g., ENFP) do you think best describes you?

18. List five ways you benefit from knowing your MBTI personality type.

19. Find out if you can take the MBTI on your campus (at the testing office, counseling center, etc.). If you can, take the test and determine your MBTI personality type. Compare the results with Problem 17.

20. Take the Keirsey Temperament Sorter II online at: www.keirsey.com/sorter/register.aspx. What temperament type are you? Find the description of that temperament type on the Keirsey website. How accurately does it describe you? Print out the description and ask someone who knows you well how on-target the description fits you.

21. Go to the Keirsey website (www.keirsey.com/sorter/register.aspx) and read the description of the Guardian Inspector (ISTJ) and the Idealist Healer (INFP). Do you think an ISTJ would make a good engineer? What about an INFP?

22. List five reasons why you should strive to improve your effectiveness in working and communicating with people who are different from you.

23. Conduct a personal assessment based on the "Attributes Model" by rating yourself on a scale of zero to ten (ten being highest) on each of the 11 attributes listed in Section 6.5. Identify the five areas in which you rate the lowest. Pick the three of those you feel are the most important. Develop a personal development plan to improve in each of these areas.

24. Evaluate yourself, on a scale of zero to ten (ten being highest), with regard to the following personal qualities:

    a. Enthusiasm
    b. Initiative
    c. Maturity
    d. Poise
    e. Integrity
    f. Flexibility
    g. Ability to work with other people

    What can you do to improve yourself in the areas with the lowest evaluations?

25. Assess the quality of your education as measured by Alexander Astin's "Student Involvement Model." Give yourself a rating of zero to ten (ten being highest) for each of the five areas listed in Section 6.5. Develop a plan for improving in any area you feel you need to improve. Implement the plan.

26. List ten types of documents that an engineer might have to write. Which of these documents do you feel qualified to write at this time?

27. Write a proposal seeking funding (e.g., from parents, private foundation, scholarship committee, etc.) to support your education. Explain how much money you need, why it's needed, and how giving you the money will ultimately benefit the *funding source*.

28. Develop a personal development plan for improving your writing skills over the next three to five years. Implement the plan.

29. Pick one of the personal development books listed at the end of Chapter 6 (Reference 2, 8, 9, 12, 13, or 15). Read the entire book (or a specific part of it) and write a critique of what you read.

30. Develop a personal development plan for improving your oral communication skills over the next three to five years in each of the following areas:

    a. One-on-one communication
    b. Group communication
    c. Formal presentations

    How do your plans differ in each area?

31. You are hired to conduct a telephone campaign to request contributions from graduates of your engineering school. Write an opening statement and follow-up statements or questions that would persuade an alum to make a contribution.

32. A common strategy for improving your vocabulary is to write down unfamiliar words from your reading, find their definition in the dictionary, and try to use them when you speak or write. Carry this strategy out for the following words which appeared in Chapter 6:

    paradigm
    conduce
    discourse
    efficacy

attribution
stereotype

33. Pick one of the achievements below:

    a. Jonas Salk developing the polio vaccine

    b. Daniel Boone blazing the wilderness trail

    c. Charles Lindbergh flying solo across the Atlantic

    d. Bill Gates creating Microsoft

    e. Mohammed Ali winning the world heavyweight boxing championship

    f. Mother Teresa ministering to the poor

    Research the achievement and write a short paper describing the type of team that was necessary to make the achievement possible.

34. Review the 16 "Principles of Teamwork" on Page 194. Rank them in order of importance. Pick your top five and write a paragraph for each principle, describing it in more detail and explaining why it is important to team effectiveness.

35. Do you seek opportunities for leadership? If not, explain in a short paper why such opportunities do not capture your interest. If, however, you do seek such opportunities, explain in a short essay which of the "Attributes of an Effective Team Leader" on Page 194 describe you and why.

36. Review the ten "Characteristics of an Effective Team Member" on Pages 195-196. Have you ever been a member of a team? If so, explain which of the characteristics best refer to you as a team member. List three characteristics that you need to improve to be a "world-class" team member. Develop a plan for making that improvement.

37. Conduct a personal assessment based on the four keys to good health listed in Section 6.8:

    a. Eat nutritionally

    b. Engage in regular aerobic exercise

    c. Get adequate sleep

    d. Avoid drugs

    Develop a personal development plan for any areas in which you rate yourself low.

38. For the next week, schedule your study time in blocks. Following each block of study time, schedule something you enjoy as a reward for doing your work. At the end of the week, evaluate how this plan worked for you.

39. Write your recollections of the messages your parents or guardians sent you about the value of a college education. Do you agree with these messages?

40. Pick a quotation that speaks to you from the 500+ motivational quotations at: www.discovery-press.com/discovery-press/studyengr/quotes.htm. Jot down the following:

    a. What does the quotation mean to you?

    b. What changes in your attitudes and/or behaviors does the quotation call for?

    Develop a plan for implementing the changes.

41. At the "Inspirational and Motivational Quotes" website: www.discovery-press.com/discovery-press/studyengr/quotes.htm, pick at least one quote from the following four individuals:

    a. Wayne Dyer
    b. Jim Rohn
    c. Zig Ziglar
    d. Peter Drucker

    Do an Internet search of each of these individuals and write a short paragraph on each describing who they are. Why do you think their wisdom is valued?

    Put each of the quotes on an index card and place the four cards prominently in your primary study area. Read and reflect on each quote daily for seven days.

42. Read the 200+ Benjamin Franklin quotes you will find at: www.discovery-press.com/discovery-press/studyengr/quotes.htm (Click on "Benjamin Franklin Quotes"). Pick your favorite quote. Write a one-page statement on why you picked the quote and what changes it calls on you to make in your life.

43. In addition to the four positive aspects of being a college student discussed in Section 6.9, list five others. Do you believe the adage that "positive attitudes bring positive results; negative attitudes bring negative results"? Explain your thinking.

CHAPTER 7
Broadening Your Education

*We are not just our behavior. We are
the person managing our behavior.*

— Kenneth Blanchard

INTRODUCTION

How do you view your education? A narrow view would be that you get an education by passing a prescribed set of courses. A quality education, however, is much more than that. This chapter will introduce you to ways you can broaden your education and in doing so be significantly better prepared for a successful career.

First, we will discuss the value of participating in student organizations and extracurricular life. The skills you can develop through this participation could be as important to your success as those you gain through your formal academic work.

Next, we will describe opportunities for you to gain practical engineering experience by participating in student design competitions, technical paper contests, design clinics, and research projects.

Then, we will discuss strategies and approaches for seeking pre-professional employment, including summer jobs, part-time jobs, and cooperative education work experiences. Through such pre-professional employment, you can gain valuable practical experience, better define your career goals, and earn money to help cover the cost of your education.

Next, we will discuss study abroad. There is perhaps nothing you could do that would be more broadening than spending a portion of your undergraduate time living and studying in another country. And by doing so, you will develop skills and amass experiences that will improve your employability with multinational companies working in today's global economy.

Finally, we will discuss opportunities for you to give something back to your educational institution. Opportunities for service can range from visiting high schools to recruit students to serving as a volunteer tutor to providing feedback to faculty so they can improve the quality of their teaching.

7.1 PARTICIPATION IN STUDENT ORGANIZATIONS

In Chapter 1 we presented two models for viewing your education. The *Employment Model* and the *Student Involvement Model* both suggested that you should participate actively in student organizations and extracurricular campus life (See Pages 25-27) The reasons were slightly different for each model, but very much related. The *Student Involvement Model* indicated that the quality of your education will be enhanced through participation in student organizations. The *Employment Model* identified two attributes that are highly valued by employers – "functions effectively on a team" and "possesses the ability to think . . . cooperatively" – that you can strengthen by involvement in student organizations.

If you do not participate in student organizations, I expect you have your own reasons. You may have never considered getting involved or you may not be aware of the benefits of such participation. On the other hand, you may have considered getting involved, but decided you don't have enough time or maybe you are shy, and therefore reluctant to join a group of people you don't know.

I hope I can persuade you to let go of these or other reasons you might have. Participation in student organizations can contribute significantly to the quality of your education. Through such participation, you can:

- Meet your social needs
- Develop your leadership and organizational skills
- Engage in professional development activities
- Receive academic support
- Participate in service activities

And, once again, these benefits are highly valued by employers

> Imagine yourself, as you near graduation, interviewing for a position with a local company. You can bet one of the questions you will be asked by the interviewer is, "Can you give me any examples of your involvement in student organizations, particularly those in which you took on leadership roles?" How do you think it will be viewed when you answer, *"Not really. I was too busy studying."*

A word of caution, though: Be selective about your involvement in such activities, since the opportunities to participate are so numerous that you could wind up neglecting your studies.

Your university could have literally hundreds of student organizations. These include recreational organizations, service organizations, social fraternities and sororities, ethnic- and gender-based organizations, and academic and professional organizations.

ENGINEERING STUDENT ORGANIZATIONS

Of the many different student organizations, the ones immediately accessible to you and offering the greatest potential for benefit are those within your engineering college. Most of these student organizations fall into one of three categories:

(1) Student chapters of discipline-specific national engineering societies

(2) National engineering honor societies

(3) Student chapters of national ethnic- and gender-based engineering organizations

STUDENT CHAPTERS OF DISCIPLINE-SPECIFIC NATIONAL ENGINEERING SOCIETIES. It is very likely that your engineering college has a student chapter corresponding to your engineering discipline. For example, if you are an electrical or computer engineering major, you could join the Institute of Electrical and Electronics Engineers (IEEE) student chapter. As a mechanical engineering major, you would want to become involved in the American Society of Mechanical Engineers (ASME) student chapter, and so forth.

There could be multiple student organizations for a particular discipline, each representing a specialization within that discipline. For example, in addition to the ASME student chapter, the mechanical engineering department at your institution could also be home to student chapters of the Society of Manufacturing Engineers (SME), Society of Automotive Engineers (SAE), Society for the Advancement of Material and Process Engineering (SAMPE), and American Society of Heating, Refrigerating and Air-Conditioning Engineers (ASHRAE).

Similarly, a civil engineering department could have, in addition to an American Society of Civil Engineers (ASCE) student chapter, student chapters of the Earthquake Engineering Research Institute (EERI), the Structural Engineers Association (SEA), the American Concrete Institute (ACI), and the Institute of Transportation Engineers (ITE).

NATIONAL ENGINEERING HONOR SOCIETIES. Many engineering disciplines also have an honor society. Examples of honor societies for specific disciplines are:

Civil Engineering - *Chi Epsilon*

Electrical and Computer Engineering - *Eta Kappa Nu*

Mechanical Engineering - *Pi Tau Sigma*

Industrial Engineering - *Alpha Pi Mu*

Chemical Engineering - *Omega Chi Epsilon*

Aeronautical and Aerospace Engineering - *Sigma Gamma Tau*

Nuclear Engineering – *Alpha Nu Sigma*

Agricultural and Biological Engineering – *Alpha Epsilon*

Bioengineering and Biomedical Engineering – *Alpha Eta Mu Beta*

Environmental Engineering – *Tau Chi Alpha*

Petroleum Engineering – *Pi Epsilon Tau*

Materials Science and Engineering - *Alpha Sigma Mu*

In addition to honor societies for each discipline, there is an honor society covering all engineering disciplines: *Tau Beta Pi*. *Tau Beta Pi* is the engineering counterpart of *Phi Beta Kappa*, the honor society for liberal arts students.

Tau Beta Pi The Engineering Honor Society

You cannot choose to join honor societies; instead you must be invited. These invitations are extended to junior or senior students who have achieved an academic record that places them in the top percentage of students in their major.

I encourage you to set a personal goal of gaining membership in *Tau Beta Pi*. Striving to be in the top 20 percent of your peers is a lofty yet achievable goal, and one that is well worth shooting for. Membership in *Tau Beta Pi* is a very prestigious honor. There is almost nothing you can put on your resume that will impress employers more than membership in *Tau Beta Pi*. To learn more about *Tau Beta Pi*, visit the society's webpage at: www.tbp.org.

ETHNIC- AND GENDER-BASED STUDENT ORGANIZATIONS. Your engineering college may also have one or more ethnic-based engineering student organizations. The most common of these organizations are:

> National Society of Black Engineers (NSBE)
>
> Society of Hispanic Professional Engineers (SHPE)
>
> Society of Mexican-American Engineers and Scientists (MAES)
>
> American-Indian Science and Engineering Society (AISES)
>
> Society of Asian Scientists & Engineers (SASE)

The purpose of these organizations is to increase the representation of these ethnic groups in the engineering profession. However, membership is not restricted, and all who are committed to the purpose of the organization are welcome.

Your campus may also have a student chapter of:

> Society of Women Engineers (SWE)

The purpose of SWE is to increase the representation of women in the engineering profession. Again, membership is open to all students.

Each of the above six ethic- and gender-based student organizations operate under the auspices of a national organization. You can learn more about these organizations by visiting their websites:

National Society of Black Engineers - www.nsbe.org

Society of Hispanic Professional Engineers - www.shpe.org

Society of Mexican-American Engineers and Scientists - www.maes-natl.org

American-Indian Society of Engineers and Scientists - www.aises.org

Society of Asian Scientists & Engineers - www.saseconnect.org

Society of Women Engineers - www.swe.org

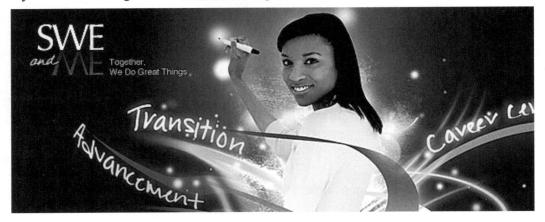

These websites also contain information about how to start a new student chapter of the organization. If a student chapter does not currently exist on your campus, you may be just the person to get one started.

ENGINEERING STUDENT COUNCIL. All of the engineering student organizations on your campus are typically organized into an engineering student council. The purpose of such an umbrella organization is to coordinate activities sponsored jointly by the student organizations, such as industry career days or events held during National Engineers Week each February.

BENEFITS OF PARTICIPATION IN STUDENT ORGANIZATIONS.

When you join an organization that is a student chapter of a national engineering society, either discipline-based or ethnic- or gender-based, (e.g., IEEE, ASME, ASCE, SHPE, SWE), by paying your dues you become a member of the national organization, and you will benefit from student activities conducted by the professional society. You will receive society publications and, in some cases, student magazines. You will be eligible to attend local, regional, and national meetings and conferences of the society. You will be eligible to compete for various awards, scholarships, and fellowships. You also will be eligible to use any career guidance or job placement services offered by the national organization.

With that said, actually your greatest benefits will come from participating in the activities of your institution's student chapter. These benefits fall into five major categories:

(1)	Social interaction
(2)	Personal development
(3)	Professional development
(4)	Academic development
(5)	Service to the college and the community

Let's explore each of these briefly.

SOCIAL INTERACTION. Participation in engineering student organizations can help you develop relationships with students who have similar backgrounds, interests, and academic and career goals as you. Close association with such students can enhance your academic success through the sharing of information and group study. The relationships you develop with fellow engineering students can continue long after your college days are over.

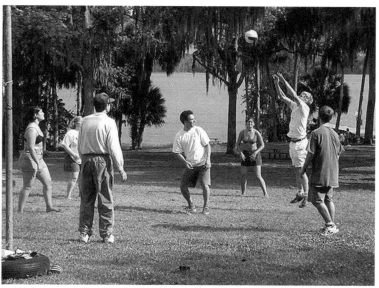

Student organizations promote this social interaction through social functions such as mixers, parties, picnics, and athletic competitions. Fund-raising activities such as car washes, raffles, jog-a-thons, and banquets also facilitate social interaction among members. Many organizations have a student lounge or study center that can greatly enhance the social environment for members.

PERSONAL DEVELOPMENT. Through participation in engineering student organizations, you can develop leadership, organizational, and interpersonal skills so important to your success as an engineering professional. You will learn from your involvement in student organizations that it is a significant challenge to get a group of people to agree on a direction and move collectively in that direction. As you learn to do this, you will acquire important skills in communicating, persuading, listening, cooperating, delegating, reporting, managing, and scheduling.

An engineering student organization can assist members in developing leadership and organizational skills by conducting leadership workshops or retreats, and by sponsoring speakers and seminars on organizational management. The greatest lessons, however, result from opportunities to practice these skills. Such opportunities can be provided to the maximum number of members by putting a committee structure in place to accomplish the various objectives appropriate to an organization (e.g., Membership Committee, Social Committee, Professional Development Committee, Academic Support Committee, High School Outreach Committee, etc.).

> ### REFLECTION
>
> Reflect on your skills in the following areas: communicating, persuading, listening, cooperating, delegating, reporting, managing, scheduling. Would you like to improve in any of these areas? Do you think participation in an engineering student organization would help you in doing so? Would taking on leadership roles help you even more?

PROFESSIONAL DEVELOPMENT. Participation in engineering student organizations can enhance your understanding of the engineering profession and the engineering work world. Much of the material presented in Chapter 2 can be brought to life through professional development activities conducted by student organizations.

Student organizations can sponsor speakers and field trips to industry. They can organize career days in which industry representatives meet with students to discuss employment need and opportunities. And they can sponsor workshops in important career development areas such as resume writing, interviewing skills, and job search strategies.

ACADEMIC DEVELOPMENT. Participation in engineering student organizations can enhance your academic performance through direct academic support activities.

Student organizations can sponsor mentor programs in which upper-division (junior and senior) students assist lower-division (freshman and sophomore) students. Organizations can arrange for volunteer tutors, organize review sessions and study groups, and maintain files of past lab reports, exams, and homework assignments. Some student organizations have their own study space to promote group study and information exchange among members. By establishing group challenge goals, such as an overall average GPA target for members, student organizations can motivate members to excel academically.

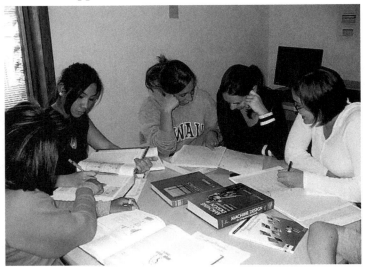

SERVICE TO COLLEGE AND COMMUNITY. Participation in engineering student organizations can provide you with a vehicle for service to the engineering college and the surrounding community.

Student organizations can sponsor visits to high schools to recruit students into engineering, raise money to establish a scholarship, sponsor activities during National Engineers Week, organize a "Meet the Dean" event, or perform other service projects that benefit the engineering college.

REFLECTION

Imagine that you are the chair of the professional development subcommittee for your engineering student organization. Your purpose is to help members learn more about engineering and to strengthen their commitment to success in becoming an engineer. What sort of activities would you propose? Seek ideas from other students you come in contact with.

PARTICIPATION IN OTHER EXTRACURRICULAR ACTIVITIES

Beyond participation in engineering student organizations, there are other extracurricular activities that can contribute to your personal development. Examples of these are writing for your campus newspaper, joining a debate club, or participating in musical or dramatic productions.

And don't forget student government – another excellent opportunity for personal growth and development. Eventually you may want to run for one of the many elected offices, but many leadership positions are filled through appointments. Visit your Associated Students office, and ask how you can become involved. ***Who knows? Maybe in a few years you'll be running for student body president!***

7.2 PARTICIPATION IN ENGINEERING PROJECTS

The quality of your education can be significantly enhanced through participation in engineering projects, contests, and competitions. Four such opportunities will be discussed here:

(1) Student design competitions
(2) Technical paper contests
(3) Design clinics
(4) Undergraduate research

STUDENT DESIGN COMPETITIONS

In recent years, the number of engineering student design competitions has grown steadily. Some of these are paper studies, but most involve the design and fabrication of an engineering device, often followed by competition against entries from other universities. Many of these competitions are sponsored by professional engineering societies listed at the end of Chapter 2. Most of the competitions involve teams of engineering students rather than individual student participation. Most have prizes, like trophies or cash awards.

Participation in one of these design competitions will give you practical, "real-world" engineering experience. You will learn to work on a complex project that is subject to strict deadlines and requires a high degree of cooperation and coordination. You will experience the type of design tradeoffs and difficult decisions that are characteristic of engineering projects. A significant investment of time will be required, but the rewards will be well worth the effort.

Your engineering college may participate in one or more of these competitions already. Check with the appropriate engineering department. If you have an interest in an event that your college does not participate in, consider taking the initiative to persuade them to do so.

The following is a representative list of these engineering student design competitions, including the name of the contest, a brief description of it, and its sponsor. To obtain more information, either go to the website of the organization listed or enter the name of the competition into a search engine such as Google.

Competition	Description	Sponsor
Reduced Gravity Education Flight Program	Propose, design, fabricate, fly, and assess a reduced-gravity experiment	**NASA Johnson Space Center** www.microgravityuniversity.jsc.nasa.gov
Chem-E-Car Competition	Design and construct a chemically-powered vehicle to carry a specified cargo a specified distance	**American Institute of Chemical Engineers** www.aiche.org
Human-Powered Helicopter Competition	$250,000 (one time only) prize for controlled flight (hover for 1 min; reach 3 m altitude; stay in 10 m sq)	**American Helicopter Society** www.vtol.org
Annual Student Design Competition	Design a rotorcraft which meets specific requirements	**American Helicopter Society** www.vtol.org
¼-Scale Tractor Design Competition	Design and build a ¼-scale tractor	**American Society of Agricultural and Biological Engineers** www.asabe.org
AGCO National Student Design Competition	Design an engineering project useful to agriculture	**American Society of Agricultural and Biological Engineers** www.asabe.org
ASABE Robotics Competition	Develop innovative robotic solutions to real-life problems in the agricultural arena	**American Society of Agricultural and Biological Engineers** www.asabe.org
Fountain Wars Design Competition	Design a fountain to be assembled in real time to perform specific functions	**American Society of Agricultural and Biological Engineers** www.asabe.org
EcoCAR2	Reduce the environmental impact of a Chevrolet Malibu	**U.S. Department of Energy/General Motors** www.ecocar2.org
Solar Splash	Design, build, and race a solar-powered boat	**Solar Splash HQ** www.solarsplash.com
Solar Decathlon	Design, build, and operate a solar-powered home	**U.S. Department of Energy** www.solardecathlon.gov
ASME Student Design Competition	Design, construct, and operate a prototype	**American Society of Mechanical Engineers** www.asme.org

Human-Powered Vehicle Challenge	Design, build, and race a streamlined bicycle	**American Society of Mechanical Engineers** www.asme.org
Student Mechanism & Robot Design Competition	Design a mechanism and robot and fabricate a prototype	**American Society of Mechanical Engineers** www.asme.org
The Great Moonbuggy Race	Design and build a vehicle that can carry two students over a half-mile simulated lunar terrain course	**NASA Marshall Space Flight Center** www.moonbuggy.msfc.nasa.gov
Concrete Canoe	Regional and national competition to design, build, and race a canoe constructed from concrete	**American Society of Civil Engineers** www.asce.org
AIAA Design Competitions	Several annual competitions to design engines, aircraft, and space satellites	**American Institute of Aeronautics and Astronautics** www.aiaa.org
ASCE/AISC National Steel Bridge Competition	Design a steel bridge that meets given specifications and build a 1:10 scale model	**ASCE and American Institute of Steel Construction (AISC)** www.aisc.org
IEEE Presidents' Change the World Competition	Demonstrate how you can change the world	**Institute of Electrical and Electronics Engineers** www.ieee.org
IEEEXtreme Global Programming Competition	Compete in a 24-hour time span to solve a set of programming problems	**Institute of Electrical and Electronics Engineers** www.ieee.org
Digilent Design Contest	Innovative digital design and microcontroller projects using Digilent products.	**Digilent, Inc.** www.digilentinc.com (Click on "Digilent Design Contest")
International Submarine Races	Design and build human-powered submarine	**Foundation for Undersea Research and Education** www.isrsubrace.org
Aero Design	Conceive, design, fabricate, and test a radio-controlled aircraft	**Society of Automotive Engineers** www.students.sae.org
Super Mileage	Design and build a vehicle powered by 3.5-HP Briggs & Stratton engine	**Society of Automotive Engineers** www.students.sae.org
Clean Snowmobile Challenge	Re-engineer an existing snowmobile	**Society of Automotive Engineers** www.students.sae.org
Formula SAE	Conceive, design, fabricate, and compete with small formula-style racing cars	**Society of Automotive Engineers** www.students.sae.org
Collegiate Inventors Competition	Submit original and inventive new ideas, processes, or technologies	**National Inventors Hall of Fame Foundation** www.invent.org/Collegiate
Baja SAE	Design, build, and race off-road vehicle powered by 10-horsepower Briggs and Stratton engine	**Society of Automotive Engineers** www.students.sae.org

ASHRAE Student Design Project Competition	Invent or optimize equipment technologies in order to solve market or societal problems	American Society of Heating, Refrigerating and Air Conditioning Engineers www.ashrae.org
Design for Direct Digital Manufacturing	Use direct digital manufacturing to develop a new and innovative product	Society of Manufacturing Engineers www.sme.org
Rube Goldberg Contest	Design and build a machine that uses the most complex process to complete a simple task	Purdue University www.purdue.edu/newsroom/rubegoldberg/index.html
American Solar Challenge	Design, build, and race a solar-powered electric vehicle	U.S. Dept of Energy www.americansolarchallenge.org
World Solar Challenge	Solar-electric vehicle race across Australian outback every two years	South Australian Motor Sports Board www.wsc.org.au

You can find a directory of over 300 national and international engineering design competitions at: www.studentcompetitions.com (Click on "Tech & Engineering").

REFLECTION

Read the brief descriptions of each of the 33 engineering student design competitions listed above. Take a few seconds right now and imagine yourself involved in the design and fabrication of a human-powered submarine, an agricultural robot, a concrete canoe, or a machine that uses the most complex process to complete a simple task (Rube Goldberg). Think through what would be involved in taking one of these projects from start to successful completion. These projects are excellent examples of the exciting challenges engineering offers.

TECHNICAL PAPER CONTESTS

Many of the professional engineering societies listed at the end of Chapter 2 sponsor technical paper contests. The contests are conducted annually and generally start with a regional competition. In most cases, regional winners progress to a national or international competition. Cash prizes are given to the top-place finishers in both the regional and national competitions. Following are some of the societies that sponsor such contests:

American Ceramic Society

American Institute of Aeronautics and Astronautics

American Institute of Chemical Engineers

American Society of Agricultural and Biological Engineers

American Society of Mechanical Engineers

American Society of Civil Engineers

Institute of Electrical and Electronics Engineers

Institute of Industrial Engineers

Society of Automotive Engineers

Society for Mining, Metallurgy, and Exploration

Society of Naval Architects and Marine Engineers

Society of Petroleum Engineers

The Minerals, Metals, and Materials Society

For more information on society-sponsored student paper contests, visit the society webpage listed at the end of Chapter 2, inquire at the engineering department office, or go see the faculty advisor of your engineering student chapter.

DESIGN CLINICS

The "design clinic" concept was pioneered at Harvey Mudd College [1] and has been adopted by a growing number of engineering colleges across the country. (Note: In 2012, three Harvey Mudd professors received the prestigious Bernard M. Gordon Prize from the National Academy of Engineering for creating and disseminating the "design clinic" concept).

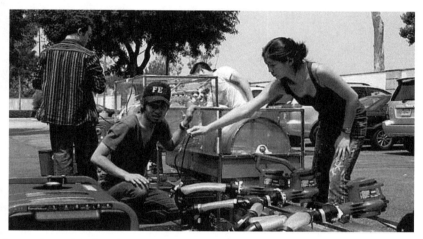

Through participation in a design clinic, you can work as part of a team of undergraduate students on a problem provided by industry. Design clinic problems could involve an engineering design problem, software development, performance testing of a product, or a theoretical study.

The design clinic work conducted by a student team is supervised by a faculty advisor. Funding is generally provided to the university by the company sponsoring the project. In some cases, students working on a design clinic may be paid. In other cases, they may receive academic credit.

Design clinics are a true "win-win" situation. Companies win by having important problems solved at a modest cost. Companies also are given the opportunity to observe the work of some of the best engineering students. As a student participating in a design clinic, you win by having the opportunity to work on a practical industry problem under the supervision of a faculty member. You also benefit from contact with practicing engineers and from getting a "foot in the door" of a company you may want to work for when you graduate.

UNDERGRADUATE RESEARCH

You can also broaden your education by working for engineering professors on their research projects. Research projects differ from design clinics in that research generally involves creating and organizing new knowledge and disseminating that knowledge through publications in technical journals and presentations at scholarly meetings.

Most likely, you only see engineering faculty as teachers. You may not realize they are also expected to conduct research. The amount of research expected of engineering professors varies from one university to the next.

Most of this research work is supported by external funding. Professors submit proposals to outside agencies requesting money to cover the costs of conducting the proposed research. One of the primary uses of the funding is to hire students to do the work. Although most of the students working on research projects are graduate students pursuing either their M.S. or Ph.D. degree, opportunities also exist for undergraduate students.

I encourage you to seek out opportunities to work on research projects during your period of undergraduate study. Openings may be listed at your institution's career center, but more likely you will have to speak to individual professors or ask about funded projects at engineering department offices.

The benefits of an undergraduate research experience can be significant. You will earn money to support the costs of your education. You will have the opportunity to work closely with an engineering professor. Since other students will probably be working on the project, you will learn how to work as part of a team.

An undergraduate research experience also gives you a chance to try out research to see if graduate school is for you. Depending on the nature of the project, you will develop your skills in specific areas such as laboratory work, computing, or engineering analysis. It is possible that you will be listed as a co-author on papers resulting from the research, and you may even have the opportunity to present the results of your work at a student research conference.

7.3 PRE-PROFESSIONAL EMPLOYMENT

An employer considering whether to hire you when you graduate would like to see that you have had previous work experience, preferably engineering-related. Engineering-related work experience not only demonstrates interest, initiative, and commitment on your part; it also provides you with references – people you have worked for who can vouch for your abilities. Prospective employers also feel that the experience you have gained will reduce the time it takes for you to become productive in their company.

BENEFITS OF PRE-PROFESSIONAL EMPLOYMENT

Pre-professional employment can benefit you in many other ways. Most obvious is that you will earn money to support the cost of your education. In addition, the process of seeking pre-professional employment can be viewed as a rehearsal for the search you will eventually conduct for a permanent job. You will develop important skills related to preparing yourself for a job search, identifying potential employers, and presenting yourself in the best light to those employers.

Pre-professional employment will enhance your professional development as well. You will gain exposure to engineering practice that will assist you in selecting your major course of study. You will gain a better understanding of the various engineering job functions. You will have an opportunity to apply your knowledge, skills, and abilities. When you return to school following your employment experience, you will have a better picture of how your academic coursework relates to the engineering work world. All of this should increase your motivation to succeed in engineering study.

TYPES OF PRE-PROFESSIONAL EMPLOYMENT

Pre-professional employment can take the form of:

Each of these is briefly discussed in the following sections.

SUMMER JOBS. Most engineering employers hire engineering students during the summer. Many employers have a formal summer job program in which they bring in a specific number of students each year.

> **Summer jobs**
>
> **Part-time jobs**
>
> **Cooperative education ("co-op") experiences**

An engineering-related summer job will not only provide you with the many benefits discussed in the previous section, it will also provide you with a welcome break from the grind of the academic year. Summer can be a time for rejuvenation. After a meaningful summer work experience, you are likely to return to school reenergized with renewed commitment.

I suggest that you set a personal goal of working in an engineering-related summer job for one or more of the summers during your undergraduate years. You need to realize, though, that student demand for summer jobs outpaces the supply. You need to adopt the positive, assertive attitude that if anyone is going to get a summer job, it's going to be you. Approaches for conducting a job search are presented in subsequent sections of this chapter.

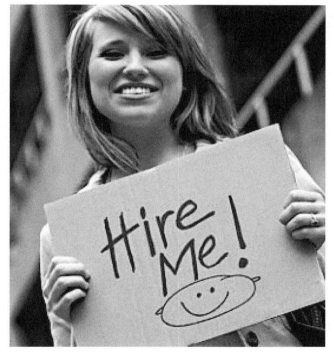

PART-TIME JOBS. You may also want to work in an engineering-related job on a part-time basis during the academic year. The availability of engineering-related jobs will depend on the location of your institution. If it is located in a major urban area, opportunities may be abundant. In contrast, if it is located in a small town, there may be no engineering employers within commuting distance.

Often students who do well in a summer job are invited to continue working on a part-time basis during the academic year. Although the employer may benefit by having you continue to work, this may not be the best situation for you. It can be flattering to be invited to continue working during the academic year, and the money may be tempting. Just make sure that you make a wise decision – one that takes your overall academic and career goals into account.

There are some tradeoffs to consider when choosing between working in a non-professional job on campus and an engineering-related job off campus. The on-campus job will take less of your time since you will not have to commute. And it will probably be easier to fit in a few hours here and there.   On the other hand, you will get more relevant experience from an engineering-related job, and the pay will probably be better.

If you do decide to work on campus, try to find a job that will complement your academic work. Working as a tutor, peer counselor, teaching assistant or grader, undergraduate research assistant, or engineering lab assistant are examples of such positions.

One final thing to remember: Full-time engineering study is a full-time commitment. You can probably work up to ten hours per week and take a full course load. If you work more than ten hours per week, you should consider reducing your course load. Recall the guidelines from Chapter 1:

Hours worked	Maximum course load
10 hrs/wk	Full load
20 hrs/wk	12 units
40 hrs/wk	8 units

Keep in mind that these are only guidelines. There are students who are able to work full-time and take a full load of courses. You will have to experiment with what works for you given your ability, background, energy level, and willingness to make personal sacrifices.

<u>COOPERATIVE EDUCATION</u>. The federal government defines *cooperative education* as:

> "a program of study at an institution of higher education under which regular students undertake academic study for specified periods of time alternating with work experience in government, industry, [or] business . . . ." [2]

The work periods can range from part-time work while engaging in part-time study (*parallel* co-op) to the more traditional "six-months-on, six-months-off" (*traditional* co-op) arrangement. The opportunity for parallel co-op is generally limited to institutions located in areas having many nearby engineering employers.

Cooperative education provides students with some distinct benefits. Among them are:

➢ Practical experience in industry
➢ Money to support college expenses
➢ A "foot in the door" in terms of seeking permanent employment upon graduation

Traditional co-op experiences will provide you with all the benefits described above, but because of the lengthier period of full-time employment, the experience gained is generally more meaningful than summer or part-time jobs. More challenging assignments can be given as progressively more experience is gained over the six-month period. The benefits of co-op are even more pronounced when the student participates in a second or third co-op experience at the same company.

The down sides of co-op are minimal. Participation in one or more traditional cooperative education work experiences will delay your graduation by up to one year. Also, some students have difficulty adjusting to their return to academia after a co-op experience. Six months of earning a good salary and having your nights and weekends free can become habit-forming.

At some institutions, co-op is a mandatory part of the engineering program. At most, however, participation is something that the student may elect. Often, students receive academic credit for the co-op assignment, but this can vary from institution to institution.

The degree of assistance that institutions provide to students seeking placement in co-op positions also varies. Places that have a mandatory co-op program will generally have a well-staffed engineering co-op office that identifies positions and matches students with them. At the other extreme, students may be virtually on their own to find co-op positions.

HOW DO YOU MEASURE UP?

Regardless of the form of pre-professional employment you seek, your competitive position will be based on three main factors:

(1) Your class level
(2) Your academic performance
(3) Your personal qualifications

As a freshman or sophomore, you will have more difficulty finding employment because companies generally prefer juniors and seniors – students closer to graduation. Companies value the opportunity to take a look at a student who will soon be a candidate for permanent employment. And junior and senior students bring a stronger technical background to their work.

But you can make up for your freshman or sophomore status by being strong in items #2 and #3. If you are a top student academically, companies will be interested in developing an early relationship with you.

The competition for top engineering students is keen. Companies are well aware that hiring you after your freshman or sophomore year will give them an advantage in hiring for a full-time position when you graduate. The question they will ask themselves is, "Are you worth the wait?"

Your personal qualifications will be a third factor in your ability to land a pre-professional employment position. The bottom-line question prospective employers will ask themselves is, ***"Will we enjoy having this student in our organization?"*** The answer will be based on an overall evaluation of your enthusiasm, initiative, communication skills, and ability to work with others. An employer will not be disappointed if you fail to solve their most pressing technical problem, but they will be very disappointed if you bring a negative, uncooperative, or unfriendly attitude to your work.

Regardless of how you measure up against the three factors discussed above, your chances of landing an engineering-related job while you are a student will depend to a great extent on how you go about your job search. <u>Effort</u> and <u>approach</u> were discussed in Chapter 1 as keys to your academic success. Similarly, effort and approach will be primary factors in your success in landing a summer job, part-time job, or co-op position. Conducting a job search not only takes considerable time and effort, it also requires that you put into practice certain strategies and approaches.

And as you start your job search, keep in mind that the opportunity exists to work abroad through international cooperative education or internships. Such experiences combine all of the benefits of pre-professional employment discussed in this section with the benefits of studying abroad discussed later in this chapter.

A job search can be divided into the following steps:

(1) Preparing yourself for the job search
(2) Identifying opportunities
(3) Applying for positions
(4) Following up on interviews

Each of these steps in the job search process is described in the following sections.

PREPARING YOURSELF FOR A JOB SEARCH

Aside from getting the best grades you possibly can and developing yourself personally using the principles presented in Chapter 6, there are specific things you need to do for a successful job search. You need to develop a resume, learn how to write cover letters, and hone your interviewing skills.

PREPARING A RESUME. A resume is a written document that lists your work experience, skills, and educational background. A resume is your main vehicle for presenting yourself to a potential employer. The central question to ask in preparing your resume is, "If you were an employer, would you want to read this resume?" Employers generally prefer smartly written one-page chronological resumes. Visual impact and appearance are extremely important. Content should include:

Identifying data (name, mailing address, telephone number, and email address)

Employment objective

Education to date

Work experience

Specialized skills

Activities and affiliations

Honors and awards

Assistance in developing your resume should be available through your career center. You can also find extensive resources on resume preparation, including formats and templates, on the Internet. One such source is the CollegeGrad.com website: www.collegegrad.com/resumes.

Click on "Resume Help" for guidance in preparing your resume. There are also a number of excellent books on resume writing [3, 4].

REFLECTION

Locate a template for your resume by either going to the CollegeGrad.com website indicated above or by googling "resume template" or "resume format." Create your resume using this template. Make a commitment to having your resume ready-to-go at all times during your period of engineering study. You never know when it will come in handy.

Most important of all, you need to tailor your resume to match the company with which you seek employment, (as this cartoon illustrates).

© Randy Glasbergen
www.glasbergen.com

"We're a big company with big ideas, and by gosh, I really like your big résumé!"

PREPARING A COVER LETTER. Whether you contact prospective employers by email, regular mail, or in person, you should always include a cover letter with your resume. And you should write a customized cover letter for each resume you send out. According to an article in the *National Business Employment Weekly* [5], if you want your cover letter "to score a direct hit in your quest for interviews" you should:

- Write to a specific person in the firm, using name and title. This should be the person who makes the hiring decision or for whom you'd work, if hired.

- In your opening paragraph, write something that demonstrates your knowledge of the organization and shows that your letter isn't a form letter.

- Communicate something about yourself that relates to the employer's needs and discusses what you can contribute.

- In your closing, ask for a meeting (do not call it an interview) and state that you will call in a few days to schedule a time.

- Limit your letter to one page, preferably printed on personalized stationery.

One additional "must": As you prepare your cover letter, pay careful attention to your organization of ideas, grammar, spelling, and the overall appearance of the letter. Many employers use cover letters to evaluate candidates' writing skills and professionalism.

The importance of your cover letter is reflected in the fact that there are whole books available to help you write a cover letter [6].

DEVELOPING YOUR INTERVIEWING SKILLS. The final area in which you need to prepare yourself is in the area of interviewing skills. Think of an interview like the *final examination* in a course. You wouldn't consider taking a final exam without extensive preparation.

If you want to gain some feedback on whether or not you are prepared to perform well in an interview, seek out a friend or fellow student and have them ask you the following questions:

How would you describe yourself?
What are your long-term career goals?
Why did you choose engineering as your major?
How would you describe your ideal job?
What was your favorite course?
Have you ever had a professor you didn't like? Explain.
What is your grade point average?
Have you taken on any leadership roles in student organizations?
Can you give any examples of working effectively on teams?
What are your greatest strengths?
What are your major shortcomings?
How would your skills meet our needs?
What have you accomplished that you are the proudest of?
What would you like to know about us?

I expect that the above exercise will convince you of the need to put significant effort into preparing yourself for interviews. You can't over-prepare for an interview. And even if you aren't asked a specific question you prepared for, you can usually find a place to work in the responses you have prepared.

You can find many more questions by googling "interview questions."

A Personal Story

When I applied for the position of Dean of Engineering, I spent an enormous amount of time preparing for the interview. I called a number of deans of engineering and asked them to tell me about the important issues facing engineering education.

I then put together a list of questions I expected to be asked and prepared written answers for each one. I practiced answering the questions on anyone I could get to ask them and sought their critique of my answers.

I think I knew more about the dean's job on the day of my interview than I did after serving in the position for many years!

In addition to practicing questions and answers, there are other ways to prepare for an interview. Learn as much as you can about the company, the job you are seeking, and the person who will be interviewing you. This task is so much easier than it used to be because of the wealth of information readily available to you on each company's website. And if the website doesn't answer all your questions, research the company at your career center or call the company directly. Also, be sure to develop a list of questions to ask the interviewer. Being inquisitive and asking good questions are sure ways to impress a person.

Aside from providing you information about a company, your campus career center can assist you in other critical ways. Most career centers offer interviewing workshops and mock interviews. In mock interviews, you are queried by a staff member assuming the role of a corporate recruiter, who then gives you valuable feedback.

Another way for you to gain insight into how well you interview is to video record yourself responding to interview questions. Video recording is a powerful tool that you should try to use.

Finally, there are many excellent references, both Internet sites and books, that can help you develop your interviewing skills [7].

IDENTIFYING EMPLOYMENT OPPORTUNITIES

There are many avenues you can take to identify pre-professional employment opportunities. Your career center is a good place to start. One of the career center's main functions is to arrange on-campus interviews. Although most of the interview opportunities will be for students seeking permanent employment, some may include interviews for summer or part-time jobs. Even if your career center doesn't provide these opportunities, it will have a list of companies that conduct on-campus interviews for engineering graduates – an excellent source of leads for you to pursue on your own. Your career center probably also maintains a bulletin board on which it posts pre-professional job listings.

Another strategy for identifying pre-professional job opportunities is to attend any job fairs or career day programs held on your campus. Try to establish personal relationships with the industry representatives there. Be friendly and sell yourself – maybe wrangle an invitation to visit their facility. It's also a good idea to carry along your resume.

NETWORKING. Networking is defined as:

> *"the process of exchanging information, contacts,*
> *and experience for professional purposes."*

In plain language, it means talking to friends, fellow students, seniors about to graduate, professors, or anyone else who might have information about job availabilities.

Other candidates for networking include practicing engineers who come to your campus to give a talk or engineering professionals you encounter at meetings of engineering societies. And don't forget alumni of your engineering program. Your college or university alumni association or engineering professors can help you identify successful alumni. But anyone, from neighbors and relatives to your doctor or people you know through church or other community affiliations, may be able to open a door for you.

Studies of successful job searches indicate networking is one of the best ways to find a job.

View everyone you know or meet as a possible lead to a job!

Remember, people enjoy helping others. If you ask people for advice, they will gladly offer it. One warning, however: People do not like being responsible for others. So don't make others feel that getting you a job is their responsibility.

One last point about networking. Don't think of it as a one-way street. Just as others can be a resource for you, you can be a resource for others. As you progress through the process of preparing and searching for jobs, you will gain valuable information that could be useful to others. Who knows? You may help a fellow student get a summer job this year, which will result in that student opening doors for you next year.

OTHER SOURCES OF EMPLOYMENT LEADS. There are many other sources of information about engineering employers. The classified ads in the newspaper can give you a clue as to who is hiring, even though the positions advertised might not be for you. Your institution's reference librarian can assist you in finding publications that list employers you can contact directly.

The North American Industry Classification System (NAICS) discussed in Chapter 2 can be used to research industries that employ engineers. First identify a specific industry (one of the 1,065 industries identified by a five- or six-digit 2012 NAICS classification code) and choose a product or service that you'd like to be involved in.

Next, google the name of the product or service to identify companies that manufacture the product or deliver the service. Go to each of these companies' webpages to learn as much as you can about them. Pick one or more of the companies and contact them about employment possibilities. Concentrating your job search on companies you find interesting and know a lot about will give you a distinct advantage.

USING THE INTERNET. The Internet provides an inexhaustible resource for tracking down job opportunities. You can seek help and direction about the best websites from your career center, or you can do it on your own. Following are some of the largest and best of the many Internet websites that could help you in your job search. (And all of these services are free!)

Monster

www.monster.com

Monster.com is the largest job search website on the Internet, claiming over a million job postings at any time and over 41 million resumes in its database. It also features employer profiles, job search and career advice, and links to other career sites.

MonsterCollege

www.college.monster.com

MonsterCollege is the most visited entry-level job search website. You'll find lots of general job search information on the MonsterCollege website, and you will be able to upload your current resume or create a new one online. Jobs are posted by type (entry-level, internships), location, and discipline.

CareerBuilder

www.careerbuilder.com

This site contains engineering job listings, a monthly newsletter, and resume and cover letter advice.

Indeed

www.indeed.com

This is a metasearch engine that uncovers job opportunities from the major job search engines and job search boards out there. You cannot submit your resume through Indeed.com, but the job search engine more than makes up for that by uncovering lots of jobs that you wouldn't find on other job search sites.

USAJobs

www.usajobs.gov

This is your gateway into the huge arena of U.S. government jobs. You'll find a wealth of resources here to help you find jobs working for Uncle Sam.

EngineerJobs

www.engineerjobs.com

This is the world's most visited job website that focuses only on engineering jobs. You can post a resume and submit an application for many jobs directly from this site.

CollegeRecruiter

www.collegerecruiter.com

This is the leading job board for college students searching for internships and recent graduates hunting for entry-level jobs.

Craigslist

www.craigslist.org

Is there anything you can't find there? Go to the website and enter a location and then under "jobs," click on "arch/engineering" to find engineering job listings.

OTHER WAYS TO USE THE INTERNET. You will also find job listings on the webpages of most of the professional engineering societies. For mechanical engineering jobs, go to jobboard.asme.org. Click on "Find Your Job" and under "Quick Job Search" enter "Mechanical Engineering."

All the major Internet search engines (e.g., Google, Bing, AltaVista, Yahoo, etc.) can be useful in exploring job opportunities. Go to any of them and enter "engineering job search."

You probably realize by now that a few hours on the Internet can provide you with more employment leads than you will ever be able to pursue. Your challenge will be to select those few that best match your needs. In the next section we tell you how to follow up on those leads.

EXERCISE

Go to www.careerbuilder.com. Under "Find Jobs," enter "engineering" and conduct a search for a listing of engineering jobs. How many are listed? Redo your search for a specific engineering discipline (e.g., civil engineering) and a specific geographical area (e.g., your state). How many jobs are listed? Scroll through the listing and identify several positions that look interesting to you. Review the detailed job description and qualifications for those positions.

APPLYING FOR POSITIONS

The most straightforward way to pursue a lead is to call the company and ask for the name and title of the individual in charge of the company's student-hiring program. Send this person a cover letter and resume. In the cover letter, state that you will follow up with a telephone call within two weeks.

Your primary goal should be to get to an interview. An interview will give you the best opportunity to sell yourself using the interviewing skills you have developed. Getting an interview is not easy, however. Industry representatives, whether they be in the human relations department or in the engineering line organization, are generally very busy. They have too many candidates for employment and too little time to evaluate applications.

If you do get an interview, follow the guidelines presented previously in the section on *Developing Your Interviewing Skills*. But you don't have to wait to be invited for an interview. You can take the initiative by arranging an "informational interview."

INFORMATIONAL INTERVIEWS. The informational interview is not a job interview. It is an information-gathering session. In a job interview, the employer is interviewing you. In an informational interview, you are interviewing the employer.

How do you arrange an informational interview? Sometimes one of your *Introduction to Engineering* course assignments will be to do an informational interview – in which case your professor will provide suggested individuals or corporations to contact. But you can, and should, arrange an informational interview on your own.

A good way to start is through networking. Perhaps through a friend or a member of your family you can get the name of an engineering manager at a local company. You then telephone that person, using the name of your friend or family member as a reference, and request 20 to 30 minutes of the person's time to learn about the company and the kind of work done there.

Although personal referrals are helpful, you can arrange informational interviews without them. Any alumnus of your engineering program would very likely be willing to meet with you. Or you can just use the fact that you are an engineering student and would like to learn more about career opportunities in engineering. Your position as a student can get you through more doors than you think. Consider the following ideas.

Student Power

Power – "the ability to influence others" – comes to people primarily from three sources: (1) their position, (2) their knowledge, and (3) their person. You probably don't realize how much power you derive from your position as a student. You are in an excellent position to influence others, and you may not even realize it.

The basis for this power is very obvious. Almost anyone you would want to influence spent many years in the very position you now hold – i.e., the position of student. And that person most likely has lots of warm, fuzzy feelings about that period of his or her life. Even more important, such people realize they owe much of their success to their education.

So when you call an engineering executive and explain, "I'm a first-year engineering student and need just 15 minutes of your time to ask you a few questions for an important project I am doing," nine times out of ten that person will agree to meet with you. Try it!

To use this newfound power, especially if you lack a personal referral, call a local engineering firm and ask to speak to the chief engineer. If you can't reach him or her, you will probably be referred to someone at a lower level. *You can then truthfully say that you were referred by the chief engineer's office and would like to meet with that person to learn more about what the company does.*

In preparing for the informational interview, make up a list of questions you plan to ask. The following are some examples:

What are your responsibilities in your current position?
What are the most satisfying aspects of your work?
What is your educational background?
Which of the courses you took in college have been most useful to you in your career?
What was your first job after graduating from college?
How did you go about getting that position?
How is your company's business picture?
What is the future hiring situation?
How important is it for engineering students to gain engineering-related work experience?
Can you advise me as to how I might get such a position that will give me that experience?

Once again, remember that people enjoy helping others and giving advice. They also like to talk about themselves. Recall the story of the coal salesman, Mr. Knaphle, in Chapter 4. By showing that you are interested in other people and want to learn from them, they will become interested in you. You may find that they offer to help you get a summer job without your even asking. Even if they don't, you can always send them an application for employment at a future date.

FOLLOWING UP ON INTERVIEWS

Whether you have a job interview or an informational interview, it is important that you follow up. Always send a thank-you letter. Few people do, so if you do, you will be remembered positively. In your letter, thank the interviewer for his or her time and interest. Be sure to mention some specific information you learned that you found particularly useful. If you are following up on a job interview, express genuine interest in the job opportunity. Conclude by leaving the door open for you to contact the person in the future.

7.4 STUDY ABROAD

In this section, we talk about study abroad. There is perhaps nothing you could do that would be more broadening than spending a portion of your undergraduate time living and studying in another country.

Studying abroad is not new. For many years, it was the purview of liberal arts students desiring to learn a foreign language through a period of immersion in a non-English speaking culture or of students just wanting to have fun "seeing the world."

But there is a growing emphasis on studying abroad within engineering education. With the onset of globalization, many engineering colleges are strongly encouraging their students to have a study-abroad experience, and some are even making such an experience mandatory.

The primary reason for this increased focus in engineering is that to a greater extent than ever before multinational corporations are seeking engineering graduates who have the experience to operate across national boundaries. In our global economy it is vital for engineers to understand cultural nuances and have the sensitivities needed to work successfully with engineers from other countries.

This priority on a global experience is reflected in the fact that one of the attributes ABET requires of all engineering graduates is

"the broad education necessary to understand the impact
of engineering solutions in a global . . . context."

You can learn something about the impact of engineering solutions in a global context in your classes, but a much better way to develop this attribute is to spend a portion of your undergraduate years living and studying abroad.

Benefits of Study Abroad

Studying abroad offers life-changing and enduring academic, career, intercultural, personal, and social benefits. A few of these benefits are:

- Developing greater self-confidence, independence, self-reliance, and maturity

- Broadening your world understanding and gaining a new perspective on the world

- Improving your cross-cultural communication skills

- Developing your ability to adapt to new and unfamiliar environments

- Building second language skills

- Making new and lasting contacts and friendships

- Distinguishing yourself from your peers in a future job search

Can You Do It?

Arranging a study-abroad experience will take time, and so you can't start too soon. Generally making all of the decisions and doing all of the planning will take about one year, so if you want to study abroad during your sophomore year you need to start planning it now. If you want to do it during your junior year or later, you can wait a bit.

To get started, locate the resources on your campus that are available to help you. Just google "<u>name of your institution</u> study abroad" and you should find lots of information about these resources. It is likely that there is a Study Abroad Office or International Programs Office with advisors to assist you. Find out whether these offices conduct periodic information sessions and attend one to help you get started.

Plenty of information about study-abroad opportunities is available on the Internet. For example, go to: www.studyabroad.com and enter "Engineering" as the subject.

There's the "good news" and the "bad news" about study abroad experiences. The good news is that there are enormous number and types of exciting opportunities. And the bad news is

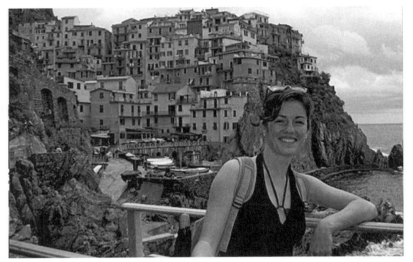

the same. There are an enormous number and types of opportunities. It can be daunting to have to choose:

- Length of time abroad
- When to go abroad
- Host country
- Method of financing
- What to study while abroad

There's an enormous world out there: many countries, many universities in each country, and many choices of coursework within each university. So it can be a difficult but exciting challenge to find the perfect situation for you. The following sections discuss each of these factors briefly.

Length of Time Abroad. Study abroad experiences vary in length. The choices you will have are:

- Few weeks (during the summer or intersession)
- Summer-long
- One semester (either fall or spring)
- Academic year (two semesters)

Certainly if you can arrange it and can afford it, the longer the experience the greater the benefits you will reap.

When to Go Abroad. Many universities require that you complete one-year of study with a certain grade point average (e.g., 2.5) before you are eligible for a study-abroad program under the institution's sponsorship. So you have a choice to go during your sophomore, junior, or senior year. Junior year is the most popular year to go abroad. This gives you more time to develop your second-language skills and more time to mature before leaving family, friends, and everything familiar. Sophomore year would generally give you more flexibility in choosing coursework that will transfer back.

<u>**Host Country.**</u> Study-abroad opportunities are available in many countries. Among the most popular for engineering study are Australia, England, New Zealand, Spain, Ireland, Italy, Germany, France, China, and Canada, but there are many others.

Here are some global destinations that are ideal for students in engineering. All are "hotbeds" of technology and engineering development and have universities with strong engineering programs.

Tokyo	Bangalore
Seoul	Singapore
London	Shanghai
Helsinki	Tel Aviv
Hong Kong	Munich

REFLECTION

How would you like to spend a semester or a whole year living and studying in one of the cities listed above? Can you imagine doing it? What do you think such an experience would do to develop you personally? Do you think it would be valuable in preparing you to work as an engineer in a global economy?

Among the factors to be considered is whether you want to go to an English-speaking country or whether you want to go to a non-English-speaking country. Even if you go to a non-English-speaking country, your instruction may still be in English.

<u>Finding a Study Abroad Program</u>

There are two categories of programs: 1) programs sponsored by your university, and 2) programs not sponsored by your university.

You will find information about programs sponsored by your university either on your Study Abroad Office website or by visiting the Study Abroad Office on your campus in person. Some advantages of university-sponsored programs:

- You remain registered at the university while abroad

- Your regular tuition is likely to cover most of the costs

- You will have an easier time getting credit for the courses you take abroad

- You are less likely to extend your time to graduation

- Any financial aid you are receiving can be used to support your study abroad

And there are programs not sponsored by your university. For these programs, you need to:

- Research available programs and apply for the program you choose individually

- Work with your registrar's office or academic department to arrange transfer credit (may be more difficult to avoid extending your time to graduation)

- Take a leave of absence from the university during the time spent abroad

- Determine whether your financial aid can be used to support your study abroad

You can find information on non-university-sponsored programs on the Internet. Some of the best sites are:

www.studyabroad.com

www.goabroad.com

www.iiepassport.org

www.transitionsabroad.com

Many countries have excellent websites for locating study abroad opportunities in that country. Two examples are:

Australia, New Zealand, and Fiji - www.australearn.org

Germany - www.daad.de/en

Financing Study Abroad. You may think you can't afford to have *a study abroad experience.* But depending on the program you choose, the cost may be less than or equal to what you are already paying. Work up a budget for your current academic year costs, work up a budget for your study-abroad experience, and compare the two. Then set about making up the difference through scholarships, financial aid, loans, or fund-raising efforts. Don't let money stand between you and the experience of a lifetime.

7.5 PUTTING SOMETHING BACK

I'm sure you have heard someone say, "I'm not going to vote. My vote doesn't really count." In one way, that view makes sense. After all, with millions of votes cast, one vote isn't really likely to make a difference. But what if everyone held the view that one vote isn't important? Since we can't afford to have everyone decide not to vote, it's not right for one person to do so.

How do you view your relationship with your university or college? Do you feel that you have something to offer your institution? Or do you feel that your contributions are not important – that what one student does cannot really make a difference? I hope you see the parallel with the importance of voting and realize that it is important to cast your vote with your university – that is, you put something back into the institution that is giving you so much.

President John F. Kennedy motivated an entire generation of young Americans when he said:

> *"Ask not what your country can do for you;*
> *ask what you can do for your country."*

Following the spirit of President Kennedy's exhortation, ask yourself, "What can I do for my institution?" Doing things that benefit your institution is bound to be a real "win-win" situation. The institution wins because what you do will make it a better place for its students, faculty, and staff. You win in two ways. You will reap direct rewards from experiences gained from what you do. And you will benefit because the quality and reputation of your institution will be improved.

Giving to your institution can and should continue throughout your lifetime. After you graduate, you will become an alumna or alumnus ("alum") of the university. As an alum, you will have the ongoing opportunity to enhance your institution through contributions of both your time and money. To some extent, the value of your education is related to the image others have of your institution. If the image of it improves, even many years after you have graduated, the value of your education will be enhanced. So whether you want it or not, you and your institution will be permanently linked.

Following are some of the ways you can put something back into your university or college, even as an undergraduate student. Doing many of the things we have already discussed, such as performing well in your classes and becoming actively involved in extracurricular activities, will by their very nature benefit your institution. But there are other ways you can contribute:

(1) Providing feedback

(2) Serving as an ambassador

(3) Helping other students

This is not a complete listing, and I'm sure you can think of other ways to give something back to your college or university. But let's consider these three suggestions for now.

PROVIDING FEEDBACK

You are your institution's primary customer. You know best whether you are getting what you need. You therefore should make every effort to let those in decision-making positions know how the institution is serving its customers. Don't restrict your feedback to negative remarks. Positive feedback can have as much, or even more, value in bringing about positive change than negative feedback.

You will have some formal opportunities to provide this kind of feedback. The best example is when you are invited to complete student opinion surveys about your professors' classroom performance. Please take these surveys seriously. They give feedback to your professors that they can use for self-improvement. Additionally, the results of the student opinion surveys are used in important decisions about tenure, promotion, and merit salary increases. Generally, these surveys consist of a series of numerical questions followed by a place for you to write narrative comments. I strongly encourage you to write detailed comments. As a professor, I found the comments far more informative and useful than the numerical results.

You will undoubtedly have other opportunities to provide feedback about your education and your institution. You may be invited to write letters of support for professors; you may receive surveys designed to measure the overall campus climate; or you may see an invitation to students to meet with the dean or department chair to give feedback. I hope you take full advantage of these and other opportunities to give feedback.

You can also give unsolicited input. Be liberal with positive feedback. As we discussed in Chapter 4, let your professors know when you like the subject or value their teaching. Tell the dean or department chairs about anything you like. People are less receptive to negative feedback, so you should be more selective with negative criticism. But if you really feel that something important is not right, don't hesitate to make an appointment to see the dean or the department chair to air your grievances. If you do, make every effort to present yourself in a tactful, respectful, and rational manner.

SERVING AS AN AMBASSADOR

You are also your institution's best ambassador. There are both formal and informal opportunities for you to serve in ambassador roles. Your college or university may have a formal *ambassadors' organization* of students who represent it at a variety of events. Such ambassadors conduct special tours; host receptions, dinners, or special events; serve as ushers; or escort distinguished visitors and alumni.

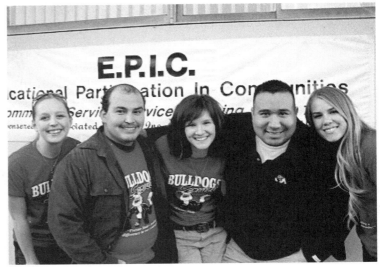

Your institution may also have a community service organization similar to the Educational Participation in Communities (EPIC) program. Through this type of organization, you can volunteer for community service assignments in schools, hospitals, community centers, and other service agencies.

You can create your own ambassador activities as well. Return to your high school or other high schools and speak to teachers and students there on behalf of your institution. Word of mouth is one of the institution's best image-builders. When you speak to anyone off campus, take the view that you are representing your college or university. Put forth the most positive perspective you are capable of. Keep in mind that any time you "bad mouth" your institution, you are diminishing the value of your education. Keep your complaints on campus and tell them to someone who can do something about them.

HELPING OTHER STUDENTS

Can you recall times when others students helped you? What did they do for you? Perhaps they pointed you toward a great teacher, provided you with information about some regulation or campus resource that really benefited you, or gave you some free tutoring that clarified a point you were stuck on.

Don't always be the one seeking help from others. Look for opportunities to help other students. Although what you have to offer will increase as you progress through the curriculum, even as a freshman you can help other students. This help can be either informal through contacts you initiate or through work as a volunteer in more structured situations. Volunteer to serve as a computer consultant in the engineering computer lab. Volunteer to work as a peer tutor in your university learning resource center. Or volunteer to work as a peer advisor in special programs for "at risk" students.

You will find that when you help others, you will get as much out of it as they do. You will develop your communication skills, increase your knowledge, and feel good about yourself for having done it.

SUMMARY

The purpose of this chapter was to introduce you to a number of activities, in addition to your formal academic work, that will broaden and enhance the quality of your education. Through participation in these activities, you will build your interpersonal communication, teamwork, organizational, and leadership skills – skills that will be important to your success in your career.

First, we described opportunities for participation in student organizations, particularly those organizations based in the engineering college, and we noted the benefits of such participation. These benefits include establishing relationships with other engineering students, developing your organizational and leadership skills, gaining valuable career information, improving your academic performance, and bolstering your self-esteem by giving something back to your engineering college or community.

Next, we discussed the value of participating in engineering projects such as student design competitions, technical paper contests, design clinics, and research projects. These activities require considerable time on your part, but the return can be enormous.

Then we discussed the value of gaining engineering-related work experience through pre-professional employment such as summer jobs, part-time jobs, and cooperative education experiences. We also presented approaches and strategies you can use in conducting successful job searches. Developing job search skills now will be invaluable to you when you seek employment as you near graduation.

Next we discussed one of the most enriching experiences you could have: studying abroad. Not only could study abroad provide you with the experience of a lifetime, you can gain a competitive edge in seeking employment in this global economy.

Finally, we described several ways you can give something back to your college or university. As your institution's most important customer, you can provide invaluable feedback. As its ambassador, you can represent it best with external constituencies. And you can be of great help to other students, just as other students have been and will continue to be of help to you.

Taking advantage of the activities described in this chapter takes initiative on your part. Unlike your formal academic work, no one will require you to do them, and no one will check up on whether you do. But the return on your investment can be even greater than the return you receive from your formal coursework. The activities outlined in this chapter truly offer opportunities for you to take responsibility for the quality of your education.

REFERENCES

1. Remer, Donald S., "Experiential Education for College Students: The Clinic – What It Is, How It Works, and How to Start One," Monograph Series of the New Liberal Arts Program, Research Foundation of the State University of New York, Stony Brook, NY, 1992.

2. Buonopane, Ralph A., "Cooperative Education – Keeping Abreast of New Technologies," *ChAPTER One*, American Institute of Chemical Engineers, May, 1990.

3. Beshara, Tony, *Unbeatable Resumes: America's Top Recruiter Reveals What REALLY Gets You Hired*, AMACOM Books, June, 2011.

4. Ireland, Susan, *The Complete Idiot's Guide to the Perfect Resume,* 5th Edition, Alpha Books, January, 2010.

5. Jackson, Tom, "Resumes, Cover Letters, and Interviews," *National Business Employment Weekly*, October, 1991.

6. Mayer, Dale, *Career Essentials: The Cover Letter,* Valley Publishing, July, 2011.

7. Deluca, Matthew, *Best Answers to the 201 Most Frequently Asked Interview Question,* 2nd Edition, McGraw-Hill, New York, NY, 2010.

PROBLEMS

1. Make a list of all the engineering student organizations at your institution. Are you an active member of one or more of these organizations? If not, join the one you are most interested in.

2. Visit the website of the national engineering society you are most interested in. Locate information on scholarships and awards given to students by the society. Share this information with fellow students in your *Introduction to Engineering* class. Determine whether you are eligible for one of the scholarships or awards, and if so, apply for it.

3. Determine whether there is a local section of the national engineering society you are most interested in. Find out when the local section holds its meetings and attend one of them. While at the meeting, try to meet as many members as you can. Ask one of them if you can visit them to conduct an informational interview.

4. Find out if the student chapter you joined is organized to accomplish the five purposes outlined in Section 7.1. If not, suggest that a committee structure be put in place to address any missing purpose (e.g., Social Committee, Professional Development Committee, Personal Development Committee, Academic Development Committee, Service Committee). Volunteer to chair one of these committees and develop a plan for the next year's activities.

5. Visit the office of your institution's student government. Arrange to meet the student body president and ask him or her whether there are any open committee assignments you could volunteer for.

6. Find out whether your engineering college has participated in any of the engineering student design competitions listed in Section 7.2. If there is one or more, which one(s)? Consider getting involved in the competition.

7. If the answer to Problem 6 is "none," pick the one you are most interested in. Contact the sponsor of the competition to obtain detailed information on the event. Try to persuade your engineering college to participate.

8. Find out if your engineering college has a design clinic program in which undergraduate students work in teams on real-world engineering problems. Find out how you can participate.

9. Find out how many full-time faculty members are in your college of engineering. Determine which ones have funded research projects. Find out how many of them employ undergraduate students to work on their research projects. Make a commitment to seek such an opportunity.

10. Determine whether your engineering college has a formal cooperative education program. (Note: The co-op program may be operated university-wide rather than by each academic unit.) If there is a formal co-op program, visit the co-op office and find out how one applies for a co-op position.

11. Visit your campus career center. Ask for a list of all companies that interview on campus for engineering graduates. Pick one of the companies and research it through its website. Write a 500-word essay about the company.

12. Do a personal assessment based on the three factors listed in the section on "How Do You Measure Up?" to find out how well you will do in competing for pre-professional employment positions. If you are not satisfied with your competitiveness, make a plan for improvement. Implement the plan.

13. Acquire one of the books on conducting job searches listed in the references for this chapter (References. 3, 4, 6, or 7). Pick three sections you are interested in and read them. Write a two-page explanation of what you learned. (Note: All of the books are available through www.amazon.com).

14. Based on the instructions given in the section on "Preparing a Resume," create your own resume. Ask several people to critique it. These could be fellow students, professors, career center staff, or practicing engineers. Revise your resume based on the input you get, and develop plans to participate in activities that would make your resume more impressive (e.g., join and participate in student organizations; find out how to earn academic awards, scholarships, or recognition; map out strategies for landing pre-professional engineering jobs; etc.). Commit to having a resume that is ready-to-go throughout your college years.

15. Write a cover letter seeking a summer job as an engineering aide with the company you selected in Problem 11. Have the cover letter critiqued by several people and revise it until you are satisfied with it. Send the letter and your resume to the company early in the spring term. Follow up on your application as explained in Section 7.3.

16. Get a friend or fellow student to ask you the questions presented in the section on "Developing Your Interviewing Skills." Have the person critique your answers.

17. Prepare a written response to each of the questions presented in the section on "Developing Your Interviewing Skills." Practice your answers and then repeat Problem 16. Did you note any difference?

18. Make up ten additional questions that you think you might be asked in an interview for a summer job. Prepare responses to those questions.

19. Use the Internet to do the following:

   a. Go to www.census.gov/naics. Click on "2012 NAICS" under "Downloads/Reference Files/Tools" to find the 24 major economic sectors.

   b. Explore one or more of the major economic sectors to identify an industry (as identified by its five- or six-digit 2012 NAICS classification code) that you are interested in.

   c. Pick a product or a service in that industry.

   d. Google the name of the product or service and identify the names of companies that make the product or deliver the service.

   e. Pick one of those companies, go to their webpage, and learn as much as you can about that company.

   f. Contact the company (by email or telephone). Find out if it has summer job positions for engineering students and, if so, how one can apply for such positions.

20. Make a list of ten companies you would like to work for in the summer using the methods outlined in the section on "Identifying Employment Opportunities." Plan a campaign to apply for a pre-professional employment position with three of them.

21. Pick one of the companies you found in Problem 19. Either through networking or through a telephone call to the chief engineer of the company, identify a person with whom you can conduct an informational interview. Arrange an interview to be conducted either in person or by telephone. Conduct the interview and write a critique discussing how the interview went. Don't forget to send a thank you letter or email after the interview.

22. Evaluate the list of questions presented in the section on informational interviews. Rate each question on a scale of one to ten. Think up five additional questions that you might ask. Rate those questions. Take the total list of 15 questions and select ten you would feel comfortable asking a practicing engineer. Put the questions in what you feel is the most logical order to ask. Use this list when you conduct any informational interviews.

23. Make up a list of questions you would ask a professor during an informational interview. Pick one of the engineering professors and arrange a 20-30 minute meeting with him or her. Write a critique of the interview.

24. Go to the CollegeMonster website: www.college.monster.com. Search under "Internships" for any part-time or summer job listings in engineering in your state. Prepare a short (two to three minutes) presentation for your *Introduction to Engineering* class on what you found there.

25. Access the EngineerJobs website: www.engineerjobs.com. Explore the jobs listed for several engineering disciplines of interest to you. Write a two-page summary about the jobs you found there.

26. Go to the "CareerBuilder" website at www.careerbuilder.com. Explore the job search resources available there. Prepare a three-minute oral presentation for your *Introduction to Engineering* classmates describing what you learned.

27. Review the benefits of a study abroad experience listed below:

Build greater self-confidence, self-reliance, and maturity
Broaden your world understanding and gain a new perspective on the world
Improve your cross-cultural communication skills
Develop your ability to adapt to new and unfamiliar environments
Acquire second language skills
Make new, lasting contacts and friendships
Distinguish yourself from your peers in future job searches

Are these benefits you would like to have? Write a short paper on where you are on two of these items and how a study-abroad experience would change you.

28. Google "<name of your institution> study abroad" to locate the website of your college or university's Study Abroad Office. Review the resources there to help you find a study abroad opportunity. Find out whether the office offers periodic information sessions. Attend one of these sessions and write a one-page paper on what you learned.

29. Go to the "StudyAbroad" website: www.studyabroad.com. Pick a city or country you might be interested in and search for programs under the subject "Engineering." Review the programs there. Prepare a 2-3 minute presentation for your *Introduction to Engineering* class on what you learned.

30. Write down ten positive features of your college or university. Rank them in order of importance. Pick ten different people (students, faculty members, department chair, dean) and tell each person about one of the features on your list. How did they respond?

31. Find out whether your institution or your engineering college has any service-oriented clubs. Write a one-page description of what one of the service organizations does. Plan to participate in the service club for at least one term during your undergraduate years.

CHAPTER 8
Orientation to the
Engineering Education System

Destiny is not a matter of chance,
it is a matter of choice.

— William Jennings Bryan

INTRODUCTION

The purpose of this chapter is to orient you to the engineering education system of which you are a part. If you are to take full advantage of your education, it is important that you understand how that educational system works. And there is an added bonus. As you better understand the engineering education system and make that system work for you, you will develop the ability to understand other systems you will encounter in the future, and you will gain the skills needed to make those systems work for you as well.

First, we will provide an overview of how engineering education is organized in the United States and how engineering programs are organized within colleges and universities.

Next, we will consider the important role community colleges play in engineering education. You may currently be a community college student, or you may have transferred from a community college to a four-year institution. Completing your first two years of engineering study at a community college can have distinct advantages. And even if the community college does not have a <u>formal</u> engineering program, you should be able to complete the majority of the lower-division engineering requirements there.

Then we will give you an overview of the engineering education system. To do this, we will use the criteria the Accreditation Board for Engineering and Technology (ABET) requires all engineering programs to meet. By understanding these criteria, you will gain insight into the key elements that comprise your engineering program: program educational objectives, student outcomes, continuous improvement, curriculum, faculty, students, facilities, and institutional support.

Next, we will examine important academic regulations, policies, and procedures in three areas: (1) academic performance, (2) enrollment, and (3) student rights. Knowing these regulations, policies, and procedures at your institution will enable you to make optimal use of the educational system.

Finally, we will consider opportunities for education beyond the B.S. degree in engineering. The benefits of pursuing an M.S. or Ph.D. degree in engineering will be discussed. Among those benefits are the advanced technical knowledge you will derive from the additional coursework and the research skills you will gain by completing a thesis or dissertation under close supervision of a faculty advisor.

We will close by exploring opportunities for post-graduate study in business administration, law, and medicine – three areas outside of engineering that many engineering graduates pursue.

8.1 ORGANIZATION OF ENGINEERING EDUCATION

According to statistics compiled by the U.S. Department of Education [1], in 2011 more than 21 million students were enrolled in 4,599 colleges and universities in the United States. Sixty-two percent were full-time students, and 38 percent studied part-time. Eighty-six percent were enrolled in undergraduate study; 14 percent were engaged in graduate and professional study. Among the 4,599 colleges and universities are 1,729 two-year institutions, 678 public four-year institutions, and 2,192 private four-year institutions.

OVERVIEW OF ENGINEERING EDUCATION IN THE U.S.

Of the 2,870 four-year colleges and universities (678 public and 2,192 private) in the U.S., only 389 (13.6 percent) offer one or more accredited engineering programs [2]. As we learned in Chapter 2 (Section 2.4), these 389 institutions offer a total of 1,885 accredited engineering programs – an average of 4.8 programs per institution. The number is not uniform, however. A few institutions offer as many as 15 different engineering programs, while others offer only one program.

Each of the nation's 1,885 accredited engineering programs is evaluated by the Accreditation Board for Engineering and Technology (ABET). Gaining ABET accreditation is extremely important. It is unlikely that any program could survive without being accredited.

To earn that accreditation, a program must meet high standards of quality for students, faculty, curriculum, administration, facilities, and institutional support and resources. Each program must also demonstrate that its graduates have acquired specific knowledge and skills; and each program must have a "continuous improvement" process in place to further develop and improve its quality. We will discuss the accreditation process in more detail in a later section.

ORGANIZATION OF THE ENGINEERING UNIT

Each engineering program is administered by a department (e.g., Department of Industrial Engineering). Generally, a department administers only one engineering program, but it is not uncommon to find two or three programs administered by a single department (e.g., Department of Mechanical and Aerospace Engineering). At the head of each department is the department chair (or department head).

The engineering departments at most institutions are organized into a "school" or "college" of engineering, headed by the dean.

Non-engineering departments may also be part of the school or college in which the engineering departments reside. Computer science, engineering technology, and industrial technology are the three most common of these. Or the engineering departments and the computer science department could be organized into a College of Engineering and Computer Science. At some small institutions, the engineering programs may be combined administratively with the mathematics and science departments to form a College of Science and Engineering.

POSITION OF THE ENGINEERING UNIT WITHIN THE UNIVERSITY

The engineering college is only one of several schools or colleges on a university campus. Other colleges typically include the College of Business, the College of Arts and Letters, the College of Natural Science, the College of Education, and the College of Health and Human Services. All of the colleges on a campus are organized into the "Academic Affairs" unit headed by the vice president or vice chancellor for academic affairs. (Often this person also carries the title of provost.)

The vice president or vice chancellor for academic affairs reports to the president or chancellor, who oversees the entire university. In addition to academic affairs, the president or chancellor is also responsible for such ancillary operations as fiscal management, facilities management, information resources management, student affairs, institutional advancement, and auxiliary services.

The organization of the academic side of the university from the president to the engineering department chairs, is shown below.

REFLECTION (FOR STUDENTS AT FOUR-YEAR INSTITUTIONS)

Do you understand the academic organization of your institution? How many ABET-accredited engineering programs does your institution offer? Do you know your department chair? Who is your dean? Who is your vice president for academic affairs? Would it be beneficial for you to know those people and for them to know you? What could you do to bring that about?

8.2 THE ROLE OF COMMUNITY COLLEGES IN ENGINEERING EDUCATION

Community colleges comprise a major part of the nation's higher education system. As previously indicated, there are 1,729 community colleges in the U.S. Forty-two percent of the nation's 18.6 million undergraduate college students are enrolled in community colleges [1].

Community colleges are a very important part of the engineering education system. Many engineering graduates started their study at a community college and then transferred to a four-year institution to complete their B.S. degree. In fact, a recent study by the National Science Foundation [3] indicated that 40 percent of engineering graduates had attended a community college at some time.

Many community colleges offer lower-division engineering or pre-engineering programs that enable students to complete all of their lower-division requirements and then transfer to a four-year institution for their upper-division engineering coursework. Some community colleges offer Associate of Science (A.S.) degrees in engineering, but only a small fraction of engineering students complete the requirements for that degree. For example, in the 2009/10 academic year, only 2,508 A.S degrees in engineering were awarded in the U.S. compared to 72,654 B.S. degrees in engineering [1].

Where community colleges do not have formal engineering programs, students can still complete about 70 to 80 percent of the lower-division engineering requirements by taking the required calculus, chemistry, physics, and lower-division general education courses – courses offered by virtually all community colleges.

ENGINEERING TECHNOLOGY

Community colleges also offer engineering technology programs. Engineering technology is a field closely related to engineering, but has a more practical focus. The difference is explained by Lawrence J. Wolf, former president of Oregon Institute of Technology [4]:

"Engineering technology draws upon the same body of knowledge as engineering but centers more heavily on the applications related to manufacturing, testing, construction, maintenance, field service, and marketing."

Although the opportunity exists for engineering technology students to transfer to four-year institutions to pursue their B.S. degree in engineering technology, the majority of engineering technology students terminate their education with the A.S. degree. This is evidenced by the fact that 31,850 A.S. degrees in engineering technology were awarded in 2009/10 by both two-year and four-year institutions; whereas, only 16,075 B.S. degrees in engineering technology were awarded in the same year by four-year institutions [1].

Some A.S. and most B.S. degree programs in engineering technology are accredited by the Accreditation Board for Engineering and Technology (ABET). A listing of these programs by discipline, degree level, and state or region can be found on the ABET website at: www.main.abet.org/aps/Accreditedprogramsearch.aspx.

An excellent source of information about engineering technology is Chapter 2, "The Field of Engineering Technology," from Stephen R. Cheshier's now out-of-print text *Studying Engineering Technology: A Blueprint for Success* [5]. Chapter 2 of that book has been updated and is available on the web at: www.discovery-press.com/discovery-press/studyent/Chap2.pdf.

ARTICULATION AND COURSE SELECTION

Community colleges with formal engineering programs generally develop *articulation agreements* with four-year institutions in their geographic area. The articulation agreements guarantee students that specific courses taken at the community college will be transferable to the four-year institution. Articulation agreements can be on a course-by-course basis, or they can apply to the full lower-division program ("2+2" articulation agreements).

Where such articulation agreements exist, they can provide you with the road map you need to plan your lower-division coursework.

Particular attention should be provided to the selection of the general education courses (humanities, social sciences, communication skills) you take at your community college. The general education requirements for an engineering degree at a four-year institution may differ from those for other majors. Unless you are careful, you might find later that you have taken courses that will not be counted among the requirements for your B.S. degree in engineering.

One additional point regarding general education courses. You may receive advice or encouragement to complete most or all of your general education requirements at your community college. I suggest that you ignore this advice and save about one-half of your general education courses to balance your technical course load during your last two years of engineering study. By doing so you'll avoid the daunting task of having to take full loads of advanced-level engineering courses during your junior and senior years.

ADVANTAGES OF STARTING AT A COMMUNITY COLLEGE

High school graduates have the choice of starting their engineering study at a community college or at a four-year institution. There are advantages and disadvantages associated with either choice. The choice usually depends on your high school record, financial situation, and personal needs. The following is a discussion of the advantages of starting at a community college.

If your record from high school does not qualify you for admission to the four-year institution of your choice, by attending a community college you, in effect, get a second chance. By building a strong academic record at the community college, you will then be able to transfer to the four-year institution of your choice. If you need to bring your math, science, and English skills up to the university level, generally a greater range of developmental courses in these areas are available at a community college than at a four-year institution.

Your time at a community college could very well provide you with the luxury of taking courses designed to give you one or more marketable skills. Examples include computer-aided-drafting, surveying, webpage design, computer programming, machine tool operation, and electronic troubleshooting. The value of taking courses in such areas is that they provide you with skills that can help you find engineering-related jobs during your tenure as a student – skills that may also come in handy later during your engineering career.

Lower cost is another advantage of attending a community college. A student who lives at home can meet his or her community college educational expenses by working as little as ten to 15 hours per week.

Finally, the community college environment lies somewhere between the warm, friendly, small-school environment you experienced in high school and the less friendly, large-school environment you will find at many major universities. Hence, you may find that a community college is a place that will provide you a more supportive learning environment in which to mature, grow, and develop before transferring to a four-year institution.

APPLICABILITY OF THIS BOOK TO COMMUNITY COLLEGE STUDENTS

If you are a community college engineering major, you will find that the concepts put forth in this book will apply directly to your situation. The first two years of engineering study at a community college are similar in virtually all regards to the first two years of engineering study at a four-year institution.

There is one exception. Various co-curricular opportunities discussed in Chapter 7 – such as research projects, engineering student organizations, engineering student design projects, and study abroad programs – are generally more available to students at four-year institutions than at community colleges. However, involvement in such activities is more likely to occur during a student's junior and senior years, so you won't miss too much.

In any case, if you are a community college student, the sections in this book that apply primarily to four-year institutions should give you a useful preview of what you can expect when you transfer to one.

REFLECTION (FOR COMMUNITY COLLEGE STUDENTS)

Are you committed to completing your lower division requirements at your community college and transferring to a four-year institution to complete your B.S. degree in engineering? Do you know what institution you plan to transfer to? Does your college have an engineering articulation agreement with that institution? Are you choosing courses based on that agreement? If you haven't decided on a university, what can you do to move closer to that decision?

8.3 THE ENGINEERING EDUCATION SYSTEM

The ABET accreditation process provides a useful framework for understanding the engineering education system. At least once every six years, a team comprised of practicing engineers and engineering educators representing the **Accreditation Board for Engineering and Technology** conducts a three-day visit to your institution to evaluate all aspects of your engineering program. The purpose of the evaluation is to ensure that the engineering program meets or exceeds specific criteria in the following eight areas discussed briefly below [6]. You can read the full criteria at: www.discovery-press.com (Click on "ABET Criteria").

CRITERION 1 - STUDENTS. This criterion addresses evaluation and monitoring of student progress, curricular and career advising, admission and transfer policies, awarding of transfer credit, and ensuring that graduates meet all graduation requirements.

CRITERION 2 - PROGRAM EDUCATIONAL OBJECTIVES. This criterion mandates that each program publish broad statements that describe what graduates are expected to attain within a few years of graduation.

CRITERION 3 -STUDENT OUTCOMES. This criterion requires each program to identify a set of student outcomes – i.e., what students are expected to know and be able to do by the time of graduation. In addition to the ABET a) – k) outcomes listed under the Attributes Model in Chapter 1 (See Page 23), programs are encouraged to articulate additional outcomes.

CRITERION 4 - CONTINUOUS IMPROVEMENT. This criterion mandates that each program have (1) a process in place for measuring whether program educational objectives and student outcomes are being attained and (2) using what is learned to continuously improve the program.

CRITERION 5 - CURRICULUM. This criterion specifies the curriculum must include one year (32 semester units) of mathematics and science, one-and-a-half years of engineering sciences and engineering design, a general education component, and a major culminating design experience.

CRITERION 6 – FACULTY. This criterion requires that faculty must be of sufficient number and have the competencies to cover all of the curricular areas of the program. Faculty must have appropriate qualifications and sufficient authority to ensure the proper guidance of the program.

CRITERION 7 - FACILITIES. This criterion mandates that adequate classrooms, offices, laboratories, and computing equipment and software be provided

CRITERION 8. INSTITUTIONAL SUPPORT. This criterion requires that adequate resources be provided for administrative and technical staff and faculty, for maintaining physical plant and infrastructure, and for acquiring and maintaining equipment. Institutional support and leadership must be adequate to ensure the quality and continuity of the program.

In addition to the eight General Criteria, each program much satisfy applicable Program Criteria, which include additional requirements in the areas of curricular topics and faculty specific to the given discipline.

Programs that meet or exceed all criteria are accredited for a six-year period. Programs with minor deficiencies will either be revisited in two years or required to write a report documenting progress in correcting the deficiencies. Serious deficiencies can result in the program being put on probation, which could lead to the loss of ABET's accreditation.

> ### REFLECTION
>
> Review the eight ABET accreditation criteria presented in the previous section. Make up a list of questions on things you would like to know more about. For example: What are the "program educational objectives" for our engineering program? Do we have "student outcomes" beyond the ABET (a) – (k) outcomes? What is meant by "design a system for sustainability"? What "modern engineering tools" will I learn to use? What course(s) address the ABET requirement for a "major design experience"? When you have the opportunity, ask your professors, your department chair, or your dean these questions.

8.4 ACADEMIC ADVISING

ABET Criterion 1 states that "Students must be advised regarding curriculum and career matters," and Criterion 6 states that "There must be sufficient faculty to accommodate adequate levels of . . . student advising and counseling . . . ."

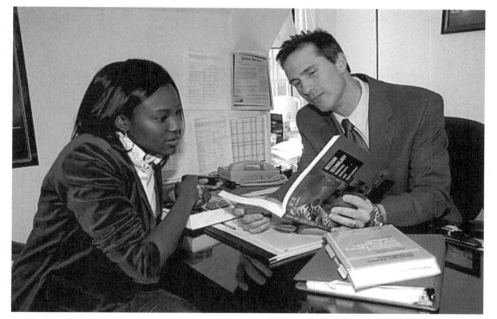

Academic advising, including both curricular and career advising, is extremely important. I hope you are studying in an engineering college in which the engineering faculty members take academic advising seriously.

You may, however, not be getting the quality of academic advising you need. Unfortunately, engineering faculty sometimes neglect their advising responsibilities in favor of the demands of teaching and research. According to Phillip C. Wankat, Professor of Engineering at Purdue University [7]:

> ***Probably the most neglected area in engineering education is advising, and certainly this is the area where students show the least satisfaction.***

Wankat's statement is borne out by my personal experiences.

> ### *Another Personal Anecdote*
>
> I recall the academic advising I received when I first began my Ph.D. program. I had been working in industry for five years and had to readjust to the demands of academic work. I was assigned an advisor who told me, "Take Dr. Johnson's course. Prove yourself by doing well in that course, and you'll have no problem from then on." Little did I know that Dr. Johnson's course was the capstone course in my field that brought together all that would be learned by completing all of the department's graduate courses. Try as hard as I could, I just couldn't handle the course. I dropped it after the midterm exam, and from then on Dr. Johnson had me pegged as a poor student.

QUALITY OF ADVISING CAN BE A PROBLEM

The absence – or minimal presence – of academic advising is not the only problem. An equally serious problem is bad advising.

One area of bad advice to look out for comes from faculty who believe that you "won't measure up" unless you graduate in four years. Such advisors will insist that you take 16-18 units, whether this is best for you or not. These faculty members can fail to account for the fact that you may be working 20 hours a week or might have been out of school for a few years and need to start slowly to work up to full speed.

Bad advice can also come from advisors who have inaccurate or out-of-date information about the curriculum or who lack information about various rules and regulations that affect your academic status. I have a constant stream of students telling me things like the following: "My advisor told me I could try out this course and drop it later." "My advisor told me it would be okay to take 20 units." "My advisor told me that ENGR 322 has been eliminated from the curriculum." Sometimes I can remedy the situation; other times I can't. Don't forget, *ignorance of the law is no excuse!*

TAKE PERSONAL RESPONSIBILITY FOR GETTING PROPER ADVISING

My recommendation to you is that you take personal responsibility for getting proper advising. After all, who suffers when you fail to be advised or get bad advice? You do!

There are several possible sources for academic advising: professors, advising staff, or other students. You certainly can be your own advisor for matters such as identifying courses you need to take, drawing up a workable schedule for a term, and so on. But you will still need sound academic and career advice from others.

To find a good advisor, first make sure you understand how the advising system works at your institution. At some institutions, advising is mandatory; at others it is optional. One department may assign students individual faculty advisors; another department may have a principal faculty advisor who advises all first-year students. Some engineering schools have advising centers where professional staff do the advising.

If you are assigned an academic advisor, whether a faculty member or professional staff member, you should meet with that person at least once each term when you plan your courses for the next term. An advising session will give you feedback on your academic performance, answer any questions you might have about academic policies or regulations, help you work out your course program for the next term, and provide you with career information.

Fellow students can be good sources of information as well. Students can be helpful in directing you to the best teachers. One warning, however. Just because one student likes a professor, doesn't mean you will. Professors are not just good or bad; they are also hard or easy. Sometimes when a student says that Professor "X" is good, he might really mean that Professor "X" is easy. I hope you will seek out professors who are good teachers, but also set high standards of performance.

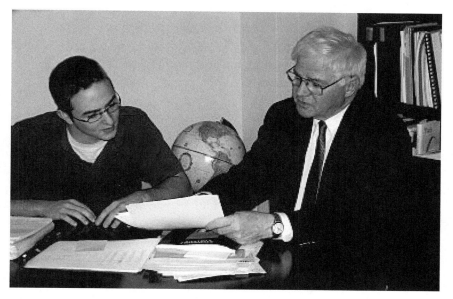

Any advice you get should be tempered with your own judgment and information you can gain from sources such as your institution's catalog, schedule of classes, website, or student handbook. These sources contain an enormous wealth of information. But you won't get that information unless you take advantage of them.

The ideal advisement arrangement is a combination of all sources. As discussed in Chapter 1, you should develop a road map that lays out the courses you plan to take each term during your undergraduate years. Share this road map regularly with your academic advisor and fellow students. Check it against the four-year curriculum outlined in your school's catalog. Based on all this input – plus your own – follow that road map or revise it whenever appropriate until you graduate.

REFLECTION

Reflect on the academic advising you have received thus far. Was it mandatory or optional? Was the advisor a professor or a student services staff member? During the advising session, did you discuss your past academic performance? Did you work out your course program for the next term? Did you discuss your long-term career goals? Were you satisfied with the quality of your advising? What could you do to ensure that you get quality advising in the future?

8.5 ACADEMIC REGULATIONS

It is also helpful to understand your institution's many academic regulations, policies, and procedures. Not knowing about some of these can hurt you; knowing about others can help you. You can find much of this information in the university catalog or on your university's webpage.

I'm not sure exactly why, but I have always been a person who has been able to get the most out of systems. Here's an example.

> ### *One Last Personal Story*
>
> As I was about to complete my B.S. degree, I realized that I could finish my M.S. degree in seven months (a summer and a semester) by taking advantage of three regulations most students had never heard of:
>
> (1) Senior-level courses beyond those needed to meet the B.S. degree requirements could be applied toward an M.S. degree.
>
> (2) Students in the last semester of undergraduate study could take two graduate-level courses toward their M.S. degree.
>
> (3) M.S. students could petition to enroll in more units than the rules allowed for a student with a full-time graduate assistantship.
>
> I knew that my GPA was marginal for admission to graduate school, so I met with the professor in charge of graduate admissions and persuaded him that my junior and senior year grades justified giving me a chance.
>
> I couldn't even start to tell you all the many ways my career has been enhanced because I stayed for those seven months and completed my M.S. degree.

I hope my story convinces you that you will benefit from understanding your institution's academic regulations, policies, and procedures. By learning them, you might be able to accomplish things that you would not otherwise even think of.

The following sections give brief overviews of important regulations, policies, and procedures that you should know about. These are divided into three categories: (1) academic performance; (2) enrollment policies; and (3) student rights.

ACADEMIC PERFORMANCE

There are a number of regulations, policies, and procedures that affect your overall academic performance. First and foremost are policies related to your grade point average and the way it is calculated. But there are other policies and procedures – such as whether you are allowed to take courses on a credit/no credit basis, how incompletes are handled, repeat grade policies, opportunities for academic renewal, and credit by examination policies – that if used optimally can help you build a strong grade point average.

GRADE POINT AVERAGE.
Your success as a student will be measured in large part by your grade point average (GPA). I can assure you from personal experience that grades <u>are</u> important. Unlike other factors that are qualitative and difficult to evaluate, your grade point average is quantitative and therefore likely to get more emphasis than it might deserve.

When I interviewed for the position of dean of engineering at California State University, Los Angeles, I was asked to submit transcripts of all my college work, and I had completed my B.S. degree 23 years before! When you interview for your first job, you may or may not be asked to submit transcripts, but you will assuredly be asked about your grade point average. If your GPA is below a certain level, some employers will eliminate you from consideration solely on that basis. Whether this practice is fair doesn't matter; it is a reality you have to face.

Most colleges and universities operate on a 4.0 grade point system as follows:

Grade Symbol	Explanation	Grade Points/Unit
A	Outstanding	4
B	Very Good	3
C	Average	2
D	Barely Passing	1
F	Failure	0

Many universities give plus and minus grades as well. This makes it easier for faculty to grade. Deciding between an *"A"* and a *"B"* or a *"B"* and a *"C"* in borderline cases can be a difficult decision for faculty. Having *"A-"* and *"B+"* or *"B-"* and *"C+"* as options makes assigning grades a lot easier, while giving a more accurate assessment of your performance.

Your total grade points are computed by summing up the product of the credit hours for each course times the grade points per unit corresponding to the letter grade you receive. Your grade point average is computed by dividing the total grade points by the total number of units taken.

One last point about your GPA. It's very important that you get off to a good start. Once you have several years behind you, it's very difficult to change your GPA. If you get your GPA off to a bad start, it'll be very difficult to raise it. But the converse is true as well. If you establish a good GPA early on, it's difficult to pull it down.

REFLECTION

One of Stephen Covey's "habits" from his excellent book *Seven Habits of Highly Effective People* is "Begin with the end in mind." As you begin your engineering study, you might want to have in mind that as you near completion of your engineering degree, you will be sitting in an interview room, perhaps at your university's career center, being interviewed for a job you would love to have. At some point in that interview, you are certain to be asked, "By the way, what's your grade point average?" How are you going to feel if you respond "3.5"? What about "3.0"? What about "2.5" or even lower? Now is the time to do the things that will make that future event a positive experience for you. This is the time to "Begin with the end in mind."

CREDIT/NO CREDIT. Many universities offer students the opportunity to take courses on a credit/no credit (CR/NC) basis. Courses taken CR/NC do not enter into the calculation of your grade point average. Generally, major requirements cannot be taken CR/NC, and the number of units that can be taken on this basis is limited. The benefit of this option, if available, is that it allows you to take courses outside of your areas of strength without the risk of lowering your GPA.

INCOMPLETES. When you are unable to complete a course for justifiable reasons (e.g., illness, family crisis, job change), you probably can request a grade of incomplete (I) from your professor. Generally, the incomplete must be made up within a certain time period. The additional time, however, provides you the opportunity to achieve a higher level of mastery in the course than if you tried to complete it in the midst of a personal crisis.

REPEAT GRADE POLICY. Your university may allow you, under specific conditions, to repeat a course and to count only the higher of the two grades you receive in your grade point average. Generally, you are only allowed to take advantage of this regulation for a limited number of courses. Some universities only allow you to repeat courses in which you have received a grade of *"D"* or *"F."* At other universities, you can even repeat a course to raise a grade of *"C"* to a *"B"* or *"A."* Check your campus regulations on this.

<u>ACADEMIC RENEWAL.</u> Your university may have a policy that allows students to remove one or more entire terms of coursework from their academic record. Generally, this can only be done under very restrictive circumstances. The policy is designed to forgive students who had one or two terms in which their academic performance was extremely low and not representative of what they are capable of. Once again, check your campus policies.

<u>CREDIT BY EXAMINATION.</u> Most universities permit students to challenge courses by examination. This is not a "free ride," however, because whatever grade you receive on the examination, including an *"F,"* is generally averaged into your grade point average.

CONSEQUENCES OF POOR ACADEMIC PERFORMANCE

Poor academic performance can lead to probation and eventual disqualification. It goes without saying that probation and disqualification should be avoided if at all possible. I have friends who flunked out of college 40 years ago and they still feel ashamed about it.

<u>PROBATION.</u> If your grades fall below a certain level, you will be placed on probation. Being placed on probation is a serious warning and indicates that unless your academic performance improves, you will be disqualified. Some universities require that students on probation receive mandatory academic advising and/or reduce the number of units they take.

<u>DISQUALIFICATION.</u> Continued poor academic performance will lead to disqualification. As previously indicated, flunking out is no fun and should be avoided at all costs. Policies for reinstatement following disqualification vary from one institution to the next. Some institutions reinstate students immediately following a first disqualification; whereas others require students to drop out of school for a period of time. If you are disqualified a second or third time, you could be permanently barred from the university.

RECOGNITION FOR GOOD ACADEMIC PERFORMANCE

The most positive recognition you can receive for your academic performance is to be granted your B.S. degree. Other positive recognitions include the Dean's List and Graduation with Honors.

<u>GRADUATION REQUIREMENTS.</u>
To graduate, you must complete all course requirements for your major with at least a 2.0 overall grade point average. Your university may also require that you have at least a 2.0 GPA in certain categories of courses such as those courses in your major field, all general education courses, and all courses attempted at your university. Other typical graduation requirements could include a time limit on courses taken and evidence

Klampkin 2012

of skills acquisition such as passing an "exit" writing proficiency exam.

DEAN'S LIST. The Dean's List is a very prestigious honor awarded each term to students who achieve a certain level of excellence. "Dean's List" status generally goes to full-time students whose grades are in the top five percent of students in their major or have achieved a certain grade point average (e.g., 3.5 GPA). Check your university's requirements, as they vary from institution to institution.

GRADUATION WITH HONORS. One of the top recognitions you can receive is to graduate with honors. There are typically three levels of honors: (1) Cum Laude (top 5%); (2) Magna cum Laude (top 3%); and (3) Summa cum Laude (top 1%). Receipt of these honors will be designated on your diploma and permanent transcript.

REFLECTION

Reflect on the various ways you can receive recognition for your academic performance. Imagine that you have just received notification that you have been placed on the Dean's List based on your academic performance for the past term. How would you feel about this? You're nearing graduation and you receive notification that you will be graduating Magna Cum Laude. How would you feel about that? It's graduation day and your family and close friends have gathered to celebrate completion of your engineering degree. How would that be?

ENROLLMENT POLICIES

Every university has a number of policies and procedures related to enrollment. These range from how you go about selecting or changing your major to how you register for your classes. Some of the most important of these policies are outlined below.

SELECTING YOUR MAJOR. The procedure by which students select their major differs among institutions. Some require students to designate an engineering discipline during their initial application process. Others will admit students as *engineering (undecided),* and students must then apply after their first or second year for admission into a specific discipline based on their academic performance.

Selecting an engineering discipline can be difficult. Your selection should be based on such factors as your aptitude, interest, and employment opportunities – factors about which you may have limited information. At universities where engineering programs have a common lower-division core for all disciplines, the decision can be postponed until the junior or even senior year. For programs that have a highly specialized curriculum requiring an early decision, you may be forced into a decision before you have adequate information.

My advice is to postpone your decision as long as your institution will permit. As you progress through the curriculum, you will be in a better position to choose because of what you learn from your coursework, pre-professional employment, and discussions

with students, professors, and practicing engineers.

CHANGING YOUR MAJOR. Don't feel that you must stay with your initial choice of major. As indicated in the previous section, you will gain insights along the way that will enable you to make more judicious decisions about your major. I started out as an electrical engineering major. I didn't like my first "Circuit Theory" course, so I changed to aeronautical engineering. Midway through a course in "Wing Design," I realized that this major was too specialized for me, so I changed to mechanical engineering. Mechanical engineering has turned out to be a great choice for me. My philosophy is summarized by the thought that:

> *If you get to a __good__ end point, then the path*
> *that took you there must have been a __good__ path.*

Don't be worried if you're not sure about what you want to do. View your college years as a chance to explore the possibilities with the purpose of finding out what you like. Take advantage of that opportunity.

One warning. You should check out the procedures for changing your major from one engineering discipline to another. Some disciplines may be easy to get into, while others may be oversubscribed and difficult to enter. Don't assume you will be able to change to whatever major you choose.

DOUBLE MAJORS. You can elect to have more than one major. Students with double majors must complete all requirements for both majors and receive two bachelor's degrees when they graduate. However, completing the graduation requirements for two majors generally takes at least one full additional year. The extra time can be reduced by choosing a second major that has a great deal of curricular overlap with the first major. For an engineering student, second majors that fit well are mathematics, physics, or computer science.

I, however, wouldn't advocate a double major. In the additional year you would spend to fulfill the requirements for the second major, you could complete a master's degree. An M.S. degree would probably be more beneficial to you than a second bachelor's degree.

MINORS. You can also elect to have a minor field. A minor can offer you most of the benefits of a second major while requiring less additional coursework. A typical minor might entail about 12 semester units (18 quarter units) of courses. The minor can be used to strengthen your preparation in an area related to your major (e.g., mathematics, physics, biology) or to gain breadth in a completely unrelated area (e.g., music, philosophy, creative writing).

REGISTRATION. The process of registering for your courses is extremely important. Through the registration process, you can ensure you get the courses you need, the best instructors, and a workable schedule. Not getting the courses you need can impede your progress, particularly in cases where the courses are prerequisites for future courses. Having the best instructors can also have a major impact on your academic success. And having a good schedule can ensure that you have adequate time for studying and other commitments.

At most institutions, registration is done online. You will generally receive a time at which you are allowed to register based on priorities established by your registrar's office. For example, the order might be new freshmen register first, followed by seniors, juniors, sophomores, and continuing freshmen.

If you are given a low-priority time slot, making it difficult to get your classes, you may be able to do something about it. Perhaps the registrar needs volunteers to work at some part of the orientation or registration process and in return allows those students to register early. Often, athletes, band members, and student government leaders are given priority registration. Research all the ways to get priority registration and see if you can qualify for one.

DROP/ADD POLICY. You should become fully versed in your institution's drop/add policy. You can probably add a course until the end of the second or third week of the term, but if you don't start attending from the beginning, you may have great difficulty catching up. Generally, you can drop a course without any penalty up to a specific date. After that, it becomes more and more difficult to do.

LEAVE OF ABSENCE/WITHDRAWAL. If you decide to leave your institution, either temporarily or permanently, be sure to follow official procedures. It is generally easy to gain approval for a leave of absence for purposes such as other professional or academic opportunities, travel or study abroad, employment related to your educational goals, field study, medical problems, or financial need. Even if you think you want to leave permanently, don't burn your bridges. Your situation may change, and you may want to return at some point in the future.

COURSE SUBSTITUTIONS. Although the engineering course requirements may seem very rigid to you, most universities have a mechanism for substituting one course for another. For example, you may want to conduct an independent study with a professor rather than taking a specific required course. Or you may want to substitute a course taught by the economics department for a required course in engineering economics. Generally, such substitutions can be made if you gain necessary approvals.

OVERLOAD POLICY. Do you know the maximum number of units you can register for? If you want to exceed that number, your college or university usually has a procedure for you to seek approval to do so. And if your GPA is high, approval will probably be granted.

CREDIT FOR COURSES AT OTHER INSTITUTIONS. You may want to take a course at a community college or other university at some point during your engineering education. Before you do so, be sure to check out your institution's policy on this. Most likely, you must get written approval in advance if you expect to receive transfer credit when you return.

STUDENT RIGHTS

There are regulations, policies, and procedures in the area of student rights that you should know about. Most universities have a "Statement of Students Rights." For example, my university puts forth the following statement to students:

1. Students have the right to reasonable **access to professional advisement** relative to all segments of their academic programs and to their career goals related to those academic programs.

2. Students have the right to **substantial instruction** in the course content at the time scheduled for class meetings except in mitigating circumstances.

3. Students have a right to expect that **their records will not be subject to unauthorized disclosure** or access.

4. Students have the right to **know about existing student record systems** and to **examine their own records**, including letters of recommendation, and to include any additional information or responses bearing on information they find objectionable.

5. Students have the right to reasonable **access to university, college, and department policies, procedures, standards, and regulations** which affect the right of a student to enroll, remain enrolled, or withdraw from any course or program of study.

6. Students have the right to **information from each professor at the first class session about the general requirements and goals of a course** in which they are enrolled, and the general criteria upon with they will be evaluated in that course.

Items #3 and #4 are dictated by the Family Educational Rights and Privacy Act (FERPA) – a law designed to protect the privacy of your educational records.

Check to see if your university or college has a similar statement of student rights. Make sure you know what your rights are!

Of course, there are also many other rights that come through the U.S. Constitution or federal or state law such as freedom of speech, freedom against discrimination, right to assemble, freedom of religion, and right to privacy. These rights may or may not be explicitly stated in your college or university's statement of student rights.

Let's look briefly at student rights in two important categories:

(1) Petitions

(2) Grievances

PETITIONS. In the previous sections, we gave examples of the many rules, regulations, policies, and procedures that are traditionally part of most educational systems. However, as the saying goes, "Every rule is made to be broken." If you find yourself constrained by a rule or regulation in a way that just doesn't make sense, you do have a recourse. Every university has a petition for waiver of regulations policy. My university catalog, for example, states:

"Students who believe that extenuating circumstances might justify the waiver of a particular regulation or requirement may file a petition at their major department/division/college office, according to established procedures, for consideration by a faculty committee."

The particular approval process can vary from case to case. Suffice it to say, if the necessary signatures can be obtained, almost anything is possible.

STUDENT GRIEVANCES. Grievances are formal complaints by students against the institution. The complaint might be about a specific instructor or administrator. Grievances generally involve an allegation by a student of unauthorized or unjustified actions that adversely affect the student's status, rights, or privileges, including but not limited to actions based on ethnicity, religion, gender, sexual orientation, age, disability, or veteran status.

You should check on your institution's student grievance policy. Generally, such policies outline specific procedures for filing grievances. At my university, for example, the process a student follows to resolve a grievance against a faculty member has five steps. First, the student is required to attempt to resolve the grievance informally with the faculty member. If this is not satisfactory, the student is then required to seek the help of the department chair to resolve the grievance informally. If that doesn't solve the problem, the student must file a formal written grievance with the department chair, who may appoint a committee to make a recommendation. The fourth step, if necessary, is to contact the school dean, who will seek the recommendation of a school-wide committee. The fifth and last step is to notify the university student grievance committee – only if the previous four steps have failed to resolve the issue.

8.6 STUDENT CONDUCT AND ETHICS

Along with rights come responsibilities. Your university has a code of conduct that delineates actions on your part that can result in disciplinary action. As an example, the following is a list of actions that warrant disciplinary action for all public universities in California [8]:

- Cheating, plagiarism, or other forms of academic dishonesty that are intended to gain unfair academic advantage

- Furnishing false information to a campus official, faculty member, or campus office

- Forgery, alteration, or misuse of a campus document, key, or identification instrument

- Misrepresentation of oneself or an organization to be an authorized agent of a campus

- Willful, material and substantial disruption or obstruction of a university-related activity or any on-campus activity

- Participating in an activity that substantially and materially disrupts the normal operations of the university or infringes on the rights of members of the university community

- Theft of property or services from the university community or misappropriation of university resources

- Unauthorized destruction, or damage to university property or other property in the university community

- Willful, material and substantial obstruction of the free flow of pedestrian or other traffic, on or leading to campus property or an off-campus university related activity

- Unauthorized entry into, presence in, use of, or misuse of campus property

- Use, possession, manufacture, or distribution of illegal drugs or drug-related paraphernalia, or the misuse of legal pharmaceutical drugs

- Use, possession, manufacture, or distribution of alcoholic beverages, or public intoxication while on campus or at a campus-related activity

- Possession or misuse of firearms or guns, replicas, ammunition, explosives, fireworks, knives, other weapons, or dangerous chemicals on campus or at a campus-related activity

- Disorderly, lewd, indecent, or obscene behavior at a campus-related activity, or directed toward a member of the campus community

- Conduct that threatens or endangers the health or safety of any person within or related to the campus community, including physical abuse, threats, intimidation, harassment, or sexual misconduct

- Hazing, or conspiracy to haze

- Misuse of computer facilities or resources including use of another's identification or password; unauthorized entry into or transfer of a file; violation of copyright laws; sending obscene, intimidating, or abusive messages; interference with the work of another member of the campus community or with campus operations

- Violation of any published campus policy, rule, regulation, or presidential order

Each of the above behaviors can bring about disciplinary sanctions including assignment of a failing grade in a course, probation, suspension, or expulsion. Many of these acts are also crimes that can result in criminal prosecution in addition to university discipline.

Of these actions, the one that occurs most often is academic dishonesty. Because of its importance, let's review what this entails.

<u>**Academic Dishonesty**</u>

During your engineering study, you will address the topic of engineering ethics. Ethics is the study of what, on a social level, is right and wrong. Engineering ethics considers how engineers should behave in different situations: what behaviors are right and what are wrong.

NSPE Code of Ethics. The preamble to the Code of Ethics of the National Society of Professional Engineers (NSPE) provides a perspective on the importance of ethics in engineering:

> "Engineering is an important and learned profession. As members of this profession, engineers are expected to exhibit the highest standards of honesty and integrity. Engineering has a direct and vital impact on the quality of life for all people. Accordingly, the services provided by engineers require honesty, impartiality, fairness, and equity, and must be dedicated to the protection of the public health, safety, and welfare. Engineers must perform under a standard of professional behavior that requires adherence to the highest principles of ethical conduct."

I would encourage you to review the NSPE *Code of Ethics for Engineers* (www.nspe.org/Ethics/CodeofEthics/index.html). The Code will give you an idea of the standards of behavior expected of engineers. Hopefully, doing so will motivate you to begin to practice these high standards during your tenure as an engineering student.

According to the Fundamental Canons of the NSPE Code of Ethics for Engineering, engineers, in the fulfillment of their professional duties, will:

1. Hold paramount the safety, health, and welfare of the public.
2. Perform services only in areas of their competence.
3. Issue public statements only in an objective and truthful manner.
4. Act for each employer or client as faithful agents or trustees.
5. Avoid deceptive acts.
6. Conduct themselves honorably, responsibly, ethically, and lawfully so as to enhance the honor, reputation, and usefulness of the profession.

Ethical Dilemmas. Ethics is a difficult subject because it is not always clear whether a certain behavior is ethical or unethical. Often engineers face dilemmas – problems for which there are no satisfactory solutions. Thus engineers might be faced with making the "least" unethical choice. As an engineering student, you also could face ethical dilemmas. Consider the following examples:

- Your professor incorrectly totaled the points on your midterm, giving you a 78 when you really only scored 58.

- A friend has been sick and asks to copy your homework that is due in a few hours.

- A fellow student notices that you have just purchased *Adobe InDesign* software and asks to borrow the CD so he can install it on his computer.

- You have a lousy professor who gives you a student opinion survey at the end of the course to evaluate his teaching. He asks you to complete it and insert it into an envelope with your final exam.

- Your professor has announced that her office hours are "Mondays and Wednesdays 10 a.m. - 12 noon." You have gone to her office during this time interval on four occasions, and she has not been there.

- The data from your laboratory experiment doesn't make any sense. Your lab partner brings you a lab report from last term and suggests that you just use the data from that report.

- Your dean invites you to be part of a group of students to meet with the chair of the visiting ABET team. The dean asks you not to say anything negative about the engineering program.

- You notice two students in your class exchange their test papers during the final exam.

- You work part-time as a student assistant in the department office. The department secretary tells you to feel free to take whatever office supplies (paper, pencils, etc.) you need from the department office supply stock room.

REFLECTION

Reflect on the ethical dilemmas just presented. In each of these situations, what alternate courses of action could you take? Do you see that the alternate courses of action reflect competing values (e.g., pleasing someone vs. doing the right thing)? What would you do in each of these situations? Note that in some cases, it is very easy to decide what's right. In other cases, it is much more difficult.

I hope this cartoon doesn't describe you.

"Lately I've been feeling ethical.
Can you prescribe something for that"

There are some areas of academic honesty for which there should be no confusion over right and wrong. These include cheating, fabrication, facilitating academic dishonesty, and plagiarism.

<u>**CHEATING.**</u>   Cheating is defined as the act of obtaining or attempting to obtain credit for academic work through the use of any dishonest, deceptive, or fraudulent means. The following examples are intended to be representative, but not all-inclusive:

<u>Examinations/Tests Administered by Faculty or the University</u>

- Copying from another student's paper
- Employing signals to obtain answers from or provide answers to others
- Stealing or arranging for the theft of an examination
- Knowingly reviewing an unauthorized copy of an examination
- Using prohibited lecture notes or textbooks during an examination
- Using crib notes during an examination
- Having someone else take an examination in your place
- Feigning illness or telling falsehoods to avoid taking an examination at the scheduled time
- Claiming falsely that you took an examination at the scheduled time
- Storing and/or accessing course subject matter in a calculator, computer or recording device, without authorization from the instructor, when such instruments are otherwise permitted to be used during an examination period
- Utilizing calculators and/or other learning aids forbidden by the instructor
- Obtaining assistance in answering questions on a take-home examination, when such action is specifically prohibited
- Attempting to use or using bribery to obtain an undeserved grade
- Changing an answer on a graded test and claiming your response to the question was incorrectly marked wrong

<u>Papers/Laboratory Reports/Homework</u>

- Copying the work of other persons in whole or in part and claiming it as your own
- Submitting a paper obtained from any source that provides research/term papers
- Paying another writer to compose a paper and claiming authorship
- Claiming an assigned share of a team report, toward which insufficient or no contribution was made
- Lying about the reason for not submitting a report on time
- Stealing another student's report and submitting it as one's own work
- Submitting the same term paper to two or more different instructors for credit in their courses without their prior permission
- Inventing, falsifying, or altering data for a research survey or laboratory experiment
- Misrepresenting the authorship of an experiment or exercise
- Depending upon others to complete laboratory assignments or homework when instructions call for independent work
- Sabotaging someone else's laboratory work or other exercise
- Fabricating bibliographic references

Cheating on any academic assignment, including coursework, comprehensive exams, or theses, is subject to discipline for academic dishonesty.

FABRICATION. Fabrication is the intentional, unauthorized falsification or invention of information or citations in an academic exercise. One example would be to make up or alter laboratory data.

FACILITATING ACADEMIC DISHONESTY. Facilitating academic dishonesty is intentionally or knowingly helping or attempting to help another to commit an act of academic dishonesty.

PLAGIARISM. Plagiarism is knowingly representing the works or ideas of another as one's own in any academic exercise. The most extreme forms of plagiarism are the use of a paper written by another person or obtained from a commercial source, or the use of passages copied word for word without attributing the passage to the writer.

8.7 GRADUATE STUDY IN ENGINEERING

Most of our discussion to this point has been directed at completing your B.S. degree in engineering. When you do complete your undergraduate work, you can go to work as a practicing engineer in industry or government or you can continue your education by working toward a graduate degree.

The graduate degree could be in engineering or in other areas such as business, law, or medicine. Several of these opportunities for continuing your education are discussed in the following sections.

BENEFITS OF GRADUATE STUDY IN ENGINEERING

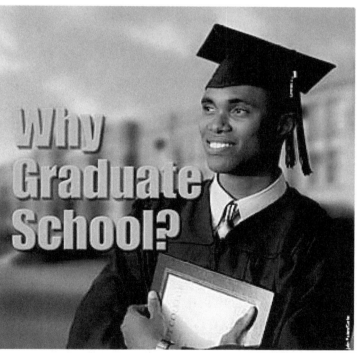

Continuing your study through an M.S. or Ph.D. degree in engineering is an invaluable investment in yourself and in your future, regardless of what you plan to do professionally. The additional years you devote to graduate study will pay off again and again throughout your career in the following ways:

- You will bolster your self-esteem and self-confidence.

- You will broaden your career choices and open doors to more challenging jobs – either in academia or in industry.

- You will increase your potential earnings over your lifetime.

- You will gain increased prestige, and others will accord you more respect.

I can't tell you how many times I have seen people's opinion of me rise when they learn that I have a Ph.D. in engineering. I can say without reservation that I would have missed much of what has been significant in my life had I not decided to seek a Ph.D.

M.S. Degree in Engineering

The first degree after the B.S. degree is the Master of Science (M.S.). Getting your master's degree takes about one year of full-time study or two years of part-time study. There are three possible options for obtaining an M.S. degree: (1) all coursework, (2) coursework plus a project, or (3) coursework plus a thesis. Some engineering colleges may offer only one option, while others offer two or even all three options.

Generally, option #1 requires about ten courses; option #2 about nine courses plus a project; and option #3 about eight courses plus a thesis. A limited number of these courses can be at the senior level with the remainder at the graduate level.

The project and thesis are similar but differ in the type and amount of work. The project can generally be completed in one term and tends to be more practical. The thesis generally takes two or more terms and involves the development of new knowledge through research.

Ph.D. Degree in Engineering

The Doctor of Philosophy (Ph.D.) is the highest educational degree in engineering. It generally takes four to five years or longer of full-time study beyond the B.S. degree. Typically, a Ph.D. program consists of about two years of coursework, culminating in a comprehensive examination (comps) covering your area of specialty. After you pass the comps, you work full-time on a research project, which will become your Ph.D. dissertation.

Normally, you would apply for admission to a Ph.D. program following completion of your M.S. degree. If you continue on at the same institution, you will generally save time. You can also complete the M.S. degree at one university and then move to another for the Ph.D. While this can be a broadening experience, it will generally extend the time required to obtain the Ph.D. degree.

At some institutions, you can be admitted directly into the Ph.D. program upon completion of your B.S. degree. In some cases you simply "pick up" the M.S. along the way with little or no additional work. At others places you may be required to take a special exam or complete a thesis to get the M.S. degree. At still other places, you can elect to skip the M.S. degree completely.

The Ph.D. degree can prepare you for a career in both industry and academia. A career as an engineering professor can provide special rewards. If you would like to know more about these, I suggest that you read "An Academic Career: It Could Be for You" [9].

Full-Time or Part-Time?

It is possible to work full-time and pursue graduate study in engineering on a part-time basis. This is the way many engineers obtain their M.S. degree. But it is much more difficult, if not impossible, to complete a Ph.D. degree on a part-time basis. Whether earning an M.S. or a Ph.D., I advise you to consider full-time graduate study if you can arrange it. Many of the benefits of graduate education come from being fully immersed in the academic environment –

concentrated study in your area of specialty, engaging in dialogue and working closely with faculty and other graduate students, and carrying out research under the close supervision of a faculty advisor.

There may be some benefits to working full-time in industry for a period after you receive your B.S. degree and then returning to full-time graduate study. You may want a break from school. You may have incurred debts you need to pay off. You may be anxious to apply what you have learned. There is, however, a potential problem with working too long in industry. You may get used to a full-time engineering salary, making it difficult to return to the more modest student life. Of course, this depends on you and your commitment to your education.

HOW WILL YOU SUPPORT YOURSELF?

Graduate study is different from undergraduate study in that ***there is a good chance you will be paid to do it.*** Any engineering graduate who has the potential to get a Ph.D. has a good chance of lining up adequate financial support for full-time graduate study. There are three kinds of financial support for graduate study: fellowships, teaching assistantships, and research assistantships. All three usually cover tuition and fees and provide a stipend for living expenses. Although fellowships and assistantships provide you much less money than full-time industrial positions, they can support you adequately to work full-time on your degree.

8.8 ENGINEERING STUDY AS PREPARATION FOR OTHER CAREERS

In Chapter 2, we made the point that an undergraduate engineering education is excellent preparation for whatever you want to do. Engineering study is particularly good preparation for graduate study in fields such as business, law, and medicine. Each of these opportunities is discussed below.

MASTER OF BUSINESS ADMINISTRATION (MBA)

One of the engineering job functions described in Chapter 2 was <u>management</u>, which typically involves either *line supervision* or *project management.* Your engineering education will not prepare you fully for these management functions. As a manager you may very well need background in economics, accounting, finance, marketing, business law, and personnel management. You will receive very little, if any, training in these subjects as part of your engineering program. The ideal academic program to give you this additional background is the Master of Business Administration (MBA).

The MBA differs from the M.S. degree in business administration. Whereas the M.S. degree in business is designed for those who did their undergraduate work in business administration, the MBA is designed for those who did their undergraduate work in other academic fields.

Admission to an MBA program does not require any prior background in business administration. Completing the MBA takes two years of full-time study. The first year is spent developing your background in accounting, economics, marketing, business law, finance, computer information systems, and management. The second year is devoted to more advanced study in these subjects, with the opportunity to specialize in one or more of them.

Admission to an MBA program is based on your undergraduate record in engineering, letters of recommendation, and scores on the Graduate Management Admission Test (GMAT). The GMAT is a national standardized test administered by the Graduation Management Admission Council (GMAC) and is offered year-round at test centers throughout the world. The test covers quantitative, verbal, and analytical writing skills.

You can learn about MBA programs and the GMAT exam by visiting the Graduate Management Admission Council (GMAC) website: www.mba.com.

Engineering students generally do very well both on the GMAT and in MBA programs. The mathematical background and strong problem-solving skills gained through an undergraduate engineering education are excellent preparation for the MBA program. If you wish to prepare further for an MBA, you should use any free electives you have as an undergraduate to take courses in economics, accounting, or behavioral science.

There are two schools of thought regarding the best path to an MBA degree. One is that you should first work for several years in a technical position to gain professional engineering experience. If you are chosen for management or decide you want to seek a management position, you would then pursue the MBA either part-time while continuing to work or by returning to school full-time.

The second school of thought is that it is better to get the MBA prior to entering the workforce. The combination of an engineering degree and MBA could lead you directly into an entry-level management position, but if it doesn't, you will be in a good position to land a management position within a short time.

One last thing regarding the MBA. If you have any thoughts of eventually starting your own company, the MBA will provide you excellent training for doing so. Being a successful entrepreneur requires competence in finance, accounting, marketing, business law, and personnel management – all areas that receive significant coverage in the MBA program.

LAW

Excellent opportunities exist for engineers in the legal profession. The primary opportunity is in patent law, where technical expertise combined with legal knowledge are essential. But other legal specialties such as environmental law and product liability law also fit well with an engineering background.

There are no specific undergraduate course requirements for law school. The most traditional pre-law majors include history, English, philosophy, political science, economics, and business, but any major including engineering is acceptable.

According to the American Bar Association (ABA) [10]:

> "The core skills and values that are essential for competent lawyering include analytic and problem-solving skills, critical reading abilities, writing skills, oral communication and listening abilities, general research skills, [and] task organization and management skills."

This would suggest that engineering study is, indeed, excellent preparation for law school, since an engineering graduate would already possess many of these skills.

There are 201 law schools approved by the ABA to offer J.D. degrees in the U.S. Admission to these approved law schools is based on undergraduate transcripts, letters of recommendation, and scores on the Law School Admissions Test (LSAT).

The LSAT is a multiple-choice exam designed to measure the following skills: reading and comprehension of complex texts with accuracy and insight; organization and management of information and the ability to draw reasonable inferences from it; the ability to reason critically; and the analysis and evaluation of the reasoning and arguments of others. The test is administered by the Law School Admission Council (LSAC) and is offered four times a year at designated centers throughout the world.

You can obtain information about law schools and the LSAT at the Law School Admission Council website: www.lsac.org.

If you are interested in law school, you should concentrate your elective undergraduate courses in history, economics, political science, and logic. Strong reading, writing, and oral communication skills are also important. Any opportunities you have to gain familiarity with legal terminology and the judicial process will also be beneficial in law school.

MEDICINE

Engineering study is also excellent preparation for medical school. Over the past several years, engineering was the third-ranked undergraduate major (behind biology and biochemistry) in the number of students who entered the nation's 137 medical schools. Perhaps more significant is that a higher percentage of engineering majors who applied were accepted and entered medical school than the average for all other undergraduate majors [11].

It is no accident that medical school admissions rates are high for engineering graduates. The logical thinking and problem-solving skills developed through engineering study have a direct carryover to the diagnostic skills practiced by physicians. The combination of engineering and medicine can lead to careers in medical research or the development of biomedical devices and equipment.

Engineering has a particular benefit over the more traditional pre-med majors such as biology, chemistry, and health science. Engineering offers students an excellent fall-back career option if they are either unable to gain admission to medical school or lose interest in a medical career.

Medical school admission requirements vary from school to school. Undergraduate course requirements include one year of biology, one year of physics, one year of English, and two years of chemistry. Experience in the health professions, extracurricular activities, and work experience are also encouraged.

Admission to medical school is based on undergraduate grades, scores on the Medical College Admission Test (MCAT), letters of recommendation, a personal statement, and a personal interview. The MCAT is a standardized, multiple-choice examination designed to assess the examinee's problem solving, critical thinking, and knowledge of science concepts and principles prerequisite to the study of medicine. The MCAT is a computer-based exam offered multiple times each year at testing sites throughout the world.

You can get information about medical schools, the MCAT exam, and the application process by visiting the Association of American Medical Colleges (AAMC) website: www.aamc.org.

The calculus, chemistry, and physics required in the engineering curriculum provide much of what is needed to prepare for the MCAT and to be admitted to medical school. Additional requirements in biology and chemistry must be taken either as elective courses or as extra courses. Biomedical engineering and chemical engineering are the two engineering disciplines that best meet the needs of a pre-med program, because additional biology and chemistry courses are part of the required curriculum.

REFLECTION

Reflect on the opportunities presented in the previous section for continuing your education beyond your B.S. degree in engineering. Do any of them appeal to you? Getting your M.S. or Ph.D. degree in engineering? Getting your MBA? Going to law school? Going to medical school? If one does, what do you need to do in the next few years to ensure that you will be prepared for and will be admitted to post-graduate study in that field?

SUMMARY

The purpose of this chapter was to orient you to the engineering education system. By understanding that system, you will be better able to make it work for you.

First, we described how the engineering college fits into the overall organization of the institution. We then discussed the role of community colleges in delivering the first two years of the engineering curriculum.

Next, we reviewed the criteria that each engineering program must meet or exceed to receive accreditation from the Accreditation Board for Engineering and Technology (ABET). The various criteria apply to eight areas: students, program educational objectives, student outcomes, continuous improvement, curriculum, faculty, facilities, and institutional support.

We also discussed academic advising, which addresses curricular and career matters. Ways were outlined for you to ensure that you receive sound academic advising, regardless of the advising system in place at your engineering college.

Then we described various academic regulations, policies, and procedures. By understanding what these entail, you can ensure that the system works <u>for</u> you, not <u>against</u> you. We also discussed the important area of student rights, including the right to petition and to file grievances.

Along with these rights comes responsibility. We therefore discussed the responsibility of students to behave ethically and honestly.

Finally, we discussed opportunities to continue your education beyond the B.S. degree. Graduate study in engineering can lead you to M.S. and Ph.D. degrees. Opportunities to seek post-graduate education in other professional fields including business administration, law, and medicine were also described.

REFERENCES

1. *Digest of Education Statistics: 2011,* National Center for Education Statistics, United States Department of Education, May, 2012. (www.nces.ed.gov/programs/digest/d11)

2. Accreditation Board for Engineering and Technology, 111 Market Place, Suite 1050, Baltimore, MD, August, 2012. (main.abet.org/aps/Accreditedprogramsearch.aspx)

3. Tsapogas, John, "The Role of Community Colleges in the Education of Recent Science and Engineering Graduates," NSF 04-315, National Science Foundation, Washington, D.C., May 2004.

4. Wolf, Lawrence J., "The Added Value of Engineering Technology," Oregon Institute of Technology, Klamath Falls, OR.

5. Cheshier, Stephen R., *Studying Engineering Technology: A Blueprint for Success*, Discovery Press, Los Angeles, CA, 1998.

6. "Criteria for Accrediting Engineering Programs: Effective for Reviews During the 2012-2013 Accreditation Cycle," Accreditation Board for Engineering and Technology, Baltimore, MD. (Available at: www.abet.org/accreditation-criteria-policies-documents)

7. Wankat, P. and Oreovicz, F., *Teaching Engineering*, McGraw-Hill, New York, NY, 1997.

8. "Title 5 – California Code of Regulations, Section 41301," California Department of Education, Sacramento, CA. (See: www.calstatela.edu/univ/stuaffrs/jao/doc/sfsc.pdf)

9. Landis, Raymond B., "An Academic Career: It Could Be for You," *Engineering Education,* July/August 1989. (Available from American Society for Engineering Education, Washington, D.C.)

10. *LSAC Official Guide to ABA-Approved Law Schools,* Law School Admission Council, 2012. (www.officialguide.lsac.org/release/OfficialGuide_Default.aspx)

11. "AAMC Data Warehouse: Applicant Matriculant File," Association of American Medical Colleges. Personal communication with Collins Mikesell (www.aamc.org/data/facts)

PROBLEMS

1. Find out the names of the people in the following positions:
   a. The chair or head of your engineering department
   b. The dean of your engineering school or college
   c. The vice chancellor or vice president for academic affairs
   d. The chancellor or president of your university

2. Locate and read the following:

   a. The mission statement of your institution
   b. Educational objectives and student outcomes of your engineering program

   Are the educational objectives and student outcomes consistent with the mission statement of the institution? Write a one-page paper discussing how they are or are not consistent.

3. Rank the (a) - (k) outcomes in Criterion 3 of the ABET Engineering Criteria 2000 in order of importance in preparing an engineering graduate for a productive career as a practicing engineer. Prepare a two-minute talk describing why you ranked the #1 item as most important.

4. Visit two engineering professors during their office hours. Ask each to rank the (a) - (k) outcomes in Criterion 3 of the ABET Engineering Criteria 2000 in order of importance. Ask each to explain why they ranked their #1 item as most important. How does their ranking compare to yours? Prepare a five-minute presentation about what you learned.

5. Identify the courses in your engineering curriculum that meet the ABET requirement for one year of mathematics and basic sciences. Do these courses equal or exceed 32 semester units (or 48 quarter units)?

6. Find out where you will learn the following computer skills in your engineering curriculum:

   a. Programming languages
   b. Word processing
   c. Computer-aided design
   d. Spread sheets
   e. Database management systems
   f. Computer graphics
   g. Data acquisition

7. Assuming you have already devised a road map showing the courses you will take to meet the requirements for your engineering major, review this plan with your academic advisor and revise it based on the feedback you receive.

8. Research the academic advising system in place in your engineering college. Write a one-page description of that advising system, including your critique of how well it works for students.

9. Does either your engineering college or your engineering department publish an *Engineering Student Handbook*? Obtain a copy and read it thoroughly. Write a one-page paper summarizing what you learned that will be of benefit to you.

10. Find out your university's regulations regarding the following academic issues:

    a. Taking courses Credit/No Credit
    b. Incompletes
    c. Repeat Grade Policy
    d. Credit by Examination
    e. Probation
    f. Disqualification
    g. Dean's List
    h. Honors at graduation

11. A student completes one semester with the following grades:

Course	Units	Grade
Math 181	5	*A*
Chemistry 150	5	*C*
Engineering 10	1	*A*
Physics 110	3	*B*

What is the student's GPA for the semester? Give the answer to two decimal places.

12. After you have completed 100 units, your overall GPA is 2.4. During the next term you take 16 units and achieve a 3.4 GPA for the term. What is your overall GPA then? If your overall GPA was 3.4 after 100 units and you take 16 units and make a 2.4 GPA for the term, what is your overall GPA then? What is the point or *message* of this exercise?

13. Determine whether your university has a S*tatement of Student Rights*. Obtain a copy and compare it to the rights discussed in Section 8.5.

14. Determine whether your university has a S*tudent Code of Conduct*. If it does, obtain a copy and review the list of actions that warrant disciplinary action.

15. Write a brief opinion as to how you would handle each of the ethical dilemmas posed in Section 8.6. Discuss your responses with at least one other student.

16. Find out how you go about changing your major from one engineering discipline to another at your institution. Are some disciplines more difficult to get into than others? Which discipline is the most difficult to get into?

17. Investigate the graduate programs available in engineering at your institution. Which engineering programs offer M.S. degrees? Which engineering programs offer Ph.D. degrees? Write down the requirements (e.g., GPA, Graduate Record Examination scores, etc.) one needs to be admitted to a graduate program in your major.

18. Consider whether you are interested in pursuing one of the three non-engineering careers discussed in Section 8.8. Locate an advisor in the area of your greatest interest (MBA, pre-law, pre-medicine) and seek additional information.

APPENDIX A – DESIGN PROJECT
Design Your Process for Becoming a "World-Class" Engineering Student

(Note: This project can be either an assignment in your *Introduction to Engineering* course or a task that you take on and complete through your own initiative.)

Engineers "design <u>products</u> or <u>processes</u> to meet desired needs." In engineering education, most of the focus is on designing products. Through this project you will design a process. You will "Design Your Process for Becoming a 'World-Class' Engineering Student." The text *Studying Engineering* will be a valuable resource in this design project.

Task:

For each of the following <u>items</u>, develop a plan that will indicate:

   a. Where a "world-class" engineering student would want to be on each item

   b. Where you are currently on each item

   c. What you need to do to move from where you are to where you would need to be to become a "world-class" engineering student

Items:

1. Set goal(s) for what you want to achieve through your engineering education (major, time to graduation, GPA, etc.) and beyond

2. Develop a strong commitment to the goal of graduating in engineering by:

   a) Clarifying what success in engineering study will do to enhance the quality of your life (rewards, benefits, opportunities, payoffs, etc)

   b) Understanding the essence of engineering (be able to articulate an answer to the question "What is engineering?)

   c) Being aware of past engineering achievements, current opportunities (academic disciplines, job functions, industry sectors) and future directions.

   d) Preparing a "road map," a term-by-term plan to guide you to graduation

   e) Other strategies identified by you.

3. Be prepared to deal with inevitable adversity

4. Do a good job of managing various aspects of your personal life including interactions with family and friends, personal finances, outside work, and commuting.

5. Change your attitudes to those appropriate to success in math/science/engineering coursework. Among those that are candidates for change are:

   a. Low self-confidence or overconfidence
   b. Reluctance to seek help
   c. Resistance to change, grow, develop, improve
   d. Tendency to procrastinate
   e. Avoidance behavior (avoid difficult or unpleasant tasks)

    f. Reluctance to study with other students

    g. Negative view toward authority figures

    h. Other negative attitudes identified by you

6. Understand teaching styles and learning styles and how to make the teaching/learning process work for you.

7. Understand and practice the concept of "metacognition" to improve your learning process by observing your learning process, feeding back to yourself what you observe, and making changes based on that feedback.

8. Change your behaviors to those appropriate to success in math/science/ engineering coursework to include at least:

    a. Devoting adequate time to studying

    b. Adopting the principle that you master the material presented in one class before the next class comes

    c. Make effective use of your peers through sharing information and group study; build productive relationships for college and beyond

    d. Make effective use of your professors both inside and outside of the classroom

    e. Prepare for lectures by reading ahead, attempting a few problems, formulating a few questions

    f. Other behaviors identified by you

9. Manage your time and tasks effectively

10. Understand the principles of teamwork and leadership and develop skills to be both an effective team member and also an effective team leader

11. Participate in co-curricular activities to good benefit

12. Understand and respect differences in learning styles and personality types and in ethnicity and gender

13. Engage in good health and wellness practices including management of stress

14. Develop a high sense of personal and professional integrity and ethical behavior

15. Become effective at getting what you want and need from the educational system by utilizing campus resources (such as advising, tutoring, job placement services, health center, etc.)

16. Add up to three additional objectives that you perceive are important for your success in engineering study

Deliverable:

Describe your plan in a 10-12 page report

APPENDIX B
21 Definitions of Engineering

The following 21 definitions of engineering from such notable sources as Count Rumford to Samuel C. Florman were complied by Harry T. Roman of East Orange, N.J., USA.

The application of science to the common purpose of life.

--Count Rumford (1799)

Engineering is the art of directing the great sources of power in nature for the use and convenience of man.

--Thomas Tredgold (1828)

It would be well if engineering were less generally thought of, and even defined, as the art of constructing. In a certain sense it is rather the art of not constructing; or, to define it rudely but not inaptly, it is the art of doing that well with one dollar which any bungler can do with two after a fashion.

--A. M. Wellington (1887)

Engineering is the art of organizing and directing men and controlling the forces and materials of nature for the benefit of the human race.

--Henry G. Stott (1907)

Engineering is the science of economy, of conserving the energy, kinetic and potential, provided and stored up by nature for the use of man. It is the business of engineering to utilize this energy to the best advantage, so that there may be the least possible waste.

--Willard A. Smith (1908)

Engineering is the conscious application of science to the problems of economic production.

--H. P. Gillette (1910)

Engineering is the art or science of utilizing, directing or instructing others in the utilization of the principles, forces, properties and substance of nature in the production, manufacture, construction, operation and use of things . . . or of means, methods, machines, devices and structures.

--Alfred W. Kiddle (1920)

Engineering is the practice of safe and economic application of the scientific laws governing the forces and materials of nature by means of organization, design and construction, for the general benefit of mankind.

--S. E. Lindsay (1920)

Engineering is an activity other than purely manual and physical work which brings about the utilization of the materials and laws of nature for the good of humanity.

--R. E. Hellmund (1929)

Engineering is the science and art of efficient dealing with materials and forces . . . it involves the most economic design and execution . . . assuring, when properly performed, the most advantageous combination of accuracy, safety, durability, speed, simplicity, efficiency, and economy possible for the conditions of design and service.

--J. A. L. Waddell, Frank W. Skinner, and H. E. Wessman (1933)

Engineering is the professional and systematic application of science to the efficient utilization of natural resources to produce wealth.

— *T. J. Hoover and J. C. L. Fish (1941)*

The activity characteristic of professional engineering is the design of structures, machines, circuits, or processes, or of combinations of these elements into systems or plants and the analysis and prediction of their performance and costs under specified working conditions.

— *M. P. O'Brien (1954)*

The ideal engineer is a composite . . . He is not a scientist, he is not a mathematician, he is not a sociologist or a writer; but he may use the knowledge and techniques of any or all of these disciplines in solving engineering problems.

— *N. W. Dougherty (1955)*

Engineers participate in the activities which make the resources of nature available in a form beneficial to man and provide systems which will perform optimally and economically.

— *L. M. K. Boelter (1957)*

The engineer is the key figure in the material progress of the world. It is his engineering that makes a reality of the potential value of science by translating scientific knowledge into tools, resources, energy and labor to bring them into the service of man . . . To make contributions of this kind the engineer requires the imagination to visualize the needs of society and to appreciate what is possible as well as the technological and broad social age understanding to bring his vision to reality.

— *Sir Eric Ashby (1958)*

The engineer has been, and is, a maker of history.

— *James Kip Finch (1960)*

Engineering is the profession in which a knowledge of the mathematical and natural sciences gained by study, experience, and practice is applied with judgment to develop ways to utilize, economically, the materials and forces of nature for the benefit of mankind.

— *Engineers Council for Professional Development (1961/1979)*

Engineering is the professional art of applying science to the optimum conversion of natural resources to the benefit of man.

— *Ralph J. Smith (1962)*

Engineering is not merely knowing and being knowledgeable, like a walking encyclopedia; engineering is not merely analysis; engineering is not merely the possession of the capacity to get elegant solutions to non-existent engineering problems; engineering is practicing the art of the organized forcing of technological change . . . Engineers operate at the interface between science and society . . .

— *Dean Gordon Brown; Massachusetts Institute of Technology (1962)*

The story of civilization is, in a sense, the story of engineering - that long and arduous struggle to make the forces of nature work for man's good.

— *L. Sprague DeCamp (1963)*

Engineering is the art or science of making practical.

— *Samuel C. Florman (1976)*

Engineers Among the World's 200 Wealthiest Individuals - March 2012 (www.forbes.com/billionaires)

Name	Position	Wealth	Major	Institution
Carlos Slim Helu	Mexican entrepreneur and businessman	$69 billion	Engineering	Universidad Nacional Autonoma de Mexico
Mikhail Gutseriev	Russian entrepreneur and businessman	$46.7 billion	Chemical Engineering	Dzhambulsky Technological Institute
Bernard Arnault	CEO Louis Vuitton Moët Hennessy	$37.5 billion	Engineering	École Polytechnique
Charles Koch	Chairman and CEO Koch Industries	$25 billion	Engineering	MIT
David H. Koch	Executive VP Koch Industries	$25 billion	Chemical Engineering	MIT
Mukesh Ambani	Chairman Reliance Industries Ltd	$22.3 billion	Chemical Engineering	University of Mumbai
Michael Bloomberg	Mayor of New York City	$22 billion	Electrical Engineering	Johns Hopkins University
Sergey Brin	Co-Founder of Google	$18.7 billion	Computer Science	University of Maryland
Larry Page	Co-Founder of Google	$18.4 billion	Computer Engineering	University of Michigan
Jeff Bezos	Founder and CEO of Amazon.com	$18.4 billion	Comp Science/ Electrical Engr.	Princeton University
Thomas Kwok	Chairman, Sun Hung Kai Properties	$18.3 billion	Civil Engineering	Imperial College
Azim Premji	Chairman and CEO Wipro Technologies	$15.9 billion	Engineering	Stanford University
Alexei Mordashov	Chairman of the Board of RUSAL	$15.3 billion	Engineering	M.V. Lomonosov Moscow State University
Vagit Alekperov	President of LUKoil Oil Company	$13.4 billion	Engineering	Azerbaijan Institute of Oil and Chemistry
Viktor Vekselberg	Chairman of Tyumen Oil (TNK)	$12.4 billion	Engineering	Moscow State University of Railroad Engineering
Antonio Ermirio de Moraes	Chair, Grupo Votorantim	$12.1 billion	Chemical Engineering	Colorado School of Mines
Leonid Mikhelson	CEO, OJSC NOVTEK	$11.9 billion	Engineering	Kuybyshev Engineering & Construction Institute
Leonard Blavatnik	Chair and President, Access Industries	$11.9 billion	Computer Science	Columbia University

Stefan Quandt	Owner Delton AG	$11.2 billion	Engineering	Technical University of Karlsruhe
Vladimir Lisin	Owner Novolipetsk steel mill	$10.7 billion	Engineering	Siberian Metallurgical Institute
Robin Li	CEO, Baidu, Inc.	$10.2 billion	Computer Science	University of Buffalo
Serge Dassault	Honorary Chairman Dassault Aviation	$9.9 billion	Aeronautical Engineering	École Polytechnique
Gennady Timchenko	Co-Founder, Gunvor Group Ltd	$9.1 billon	Elec/Mech Engineering	Mechanical Institute of St. Petersburg
Hansjörg Wyss	President, Synthes	$8.1 billion	Civil Engineering	Swiss Federal Institute of Technology Zurich
Eric Schmidt	CEO, Google	$6.9 billion	Electrical Engineering	Princeton University
Pierre Omidyar	Founder/Chairman eBay	$6.7 billion	Computer Science	Tufts University
Curt Engelhorn	Pharmaceuticals	$6.6 billion	Chemical Engineering	University of Texas Austin
Kushal Pal Singh	Chairman and CEO of DLF Limited	$6.4 billion	Aeronautical Engineering	Meerut College/United Kingdom
Bidzina Ivanishvili	Georgian businessman and politician.	$6.4 billion	Engineering	Tbilisi State University
Jeffrey Hildebrand	Chairman, and CEO. Hilcorp Energy Company	$5.3 billion	Petroleum Engineering	University of Texas Austin

APPENDIX D
Greatest Engineering Achievements of the 20th Century

#20 - HIGH PERFORMANCE MATERIALS

From the building blocks of iron and steel to the latest advances in polymers, ceramics, and composites, the 20th century has seen a revolution in materials. Engineers have tailored and enhanced material properties for uses in thousands of applications.

#19 - NUCLEAR TECHNOLOGIES

The harnessing of the atom changed the nature of war forever and astounded the world with its awesome power. Nuclear technologies also gave us a new source of electric power and new capabilities in medical research and imaging.

#18 - LASER AND FIBER OPTICS

Pulses of light from lasers are used in industrial tools, surgical devices, satellites, and other products. In communications, highly pure glass fibers now provide the infrastructure to carry information via laser-produced light – a revolutionary technical achievement. Today, a single fiber-optic cable can transmit tens of millions of phone calls, data files, and video images.

#17 - PETROLEUM AND GAS TECHNOLOGIES

Petroleum has been a critical component of 20th century life, providing fuel for cars, homes, and industries. Petrochemicals are used in products ranging from aspirin to zippers. Spurred on by engineering advances in oil exploration and processing, petroleum products have had an enormous impact on world economies, people, and politics.

#16 - HEALTH TECHNOLOGIES

Advances in 20th century medical technology have been astounding. Medical professionals now have an arsenal of diagnostic and treatment equipment at their disposal. Artificial organs, replacement joints, imaging technologies, and bio-materials are but a few of the engineered products that improve the quality of life for millions.

#15 - HOUSEHOLD APPLIANCES

Engineering innovation produced a wide variety of devices, including electric ranges, vacuum cleaners, dishwashers, and dryers. These and other products give us more free time, enable more people to work outside the home, and contribute significantly to our economy.

#14 - IMAGING TECHNOLOGIES

From tiny atoms to distant galaxies, imaging technologies have expanded the reach of our vision. Probing the human body, mapping ocean floors, tracking weather patterns – all are the result of engineering advances in imaging technologies.

#13 - INTERNET

The Internet is changing business practices, educational pursuits, and personal communications. By providing global access to news, commerce, and vast stores of information, the Internet brings people together globally while adding convenience and efficiency to our lives.

#12 - SPACE EXPLORATION

From early test rockets to sophisticated satellites, the human expansion into space is perhaps the most amazing engineering feat of the 20th century. The development of spacecraft has thrilled the world, expanded our knowledge base, and improved our capabilities. Thousands of useful products and services have resulted from the space program, including medical devices, improved weather forecasting, and wireless communications.

#11 - INTERSTATE HIGHWAYS

Highways provide one of our most cherished assets – the freedom of personal mobility. Thousands of engineers built the roads, bridges, and tunnels that connect our communities, enable goods and services to reach remote areas, encourage growth, and facilitate commerce.

#10 - AIR CONDITIONING AND REFRIGERATION

Air conditioning and refrigeration changed life immensely in the 20th century. Dozens of engineering innovations made it possible to transport and store fresh foods, for people to live and work comfortably in sweltering climates, and to create stable environments for the sensitive components that underlie today's information technology economy.

#9 - TELEPHONE

The telephone is a cornerstone of modern life. Nearly instant connections – between friends, families, businesses, and nations – enable communications that enhance our lives, industries, and economies. With remarkable innovations, engineers have brought us from copper wire to fiber optics, from switchboards to satellites, and from party lines to the Internet.

#8 - COMPUTERS

The computer has transformed businesses and lives around the world by increasing productivity and opening access to vast amounts of knowledge. Computers have relieved the drudgery of routine daily tasks, and brought new ways to handle complex ones. Engineering ingenuity fueled this revolution, and continues to make computers faster, more powerful, and more affordable.

#7 - AGRICULTURAL MECHANIZATION

The machinery of farms – tractors, cultivators, combines, and hundreds of others – dramatically increased farm efficiency and productivity in the 20th century. At the start of the century, four U.S. farmers could feed about ten people. By the end, with the help of engineering innovation, a single farmer could feed more than 100 people.

#6 - RADIO AND TELEVISION

Radio and television were major agents of social change in the 20th century, opening windows to other lives, to remote areas of the world, and to history in the making. From wireless telegraph to today's advanced satellite systems, engineers have developed remarkable technologies that inform and entertain millions every day.

#5 - ELECTRONICS

Electronics provide the basis for countless innovations – CD players, TVs, and computers, to name a few. From vacuum tubes to transistors, to integrated circuits, engineers have made electronics smaller, more powerful, and more efficient, paving the way for products that have improved the quality and convenience of modern life.

#4 - SAFE AND ABUNDANT WATER

The availability of safe and abundant water changed the way Americans lived and died during the last century. In the early 1900s, waterborne diseases like typhoid fever and cholera killed tens-of-thousands of people annually, and dysentery and diarrhea, the most common waterborne diseases, were the third largest cause of death. By the 1940s, however, water treatment and distribution systems devised by engineers had almost totally eliminated these diseases in American and other developed nations. They also brought water to vast tracts of land that would otherwise have been uninhabitable.

#3 - AIRPLANE

Modern air travel transports goods and people quickly around the globe, facilitating our personal, cultural, and commercial interaction. Engineering innovation – from the Wright brothers' airplane to today's supersonic jets – have made it all possible.

#2 - AUTOMOBILE

The automobile may be the ultimate symbol of personal freedom. It's also the world's major transporter of people and goods, and a strong source of economic growth and stability. From early Tin Lizzies to today's sleek sedans, the automobile is a showcase of 20th century engineering ingenuity, with countless innovations made in design, production, and safety.

#1 - ELECTRIFICATION

Electrification powers almost every pursuit and enterprise in modern society. It has literally lighted the world and impacted countless areas of daily life, including food production and processing, air conditioning and heating, refrigeration, entertainment, transportation, communication, health care, and computers. Thousands of engineers made it happen, with innovative work in fuel sources, power generating techniques, and transmission grids.

APPENDIX E
Description of Engineering Disciplines

The following sections describe 25 engineering disciplines. More extensive information is provided for the eight largest disciplines (in terms of number of graduates), while briefer descriptions are given for smaller disciplines.

E.1 ELECTRICAL ENGINEERING

Electrical engineering (including computer engineering) is the largest of all engineering disciplines. According to U.S. Department of Labor statistics, of the 1.5 million engineers working with the occupational title of "engineer" in the U.S. in 2011, 362,550 (24.6 percent) were electrical and computer engineers [1].

Electrical engineers are concerned with electrical devices and systems and with the use of electrical energy. Virtually every industry utilizes electrical engineers, so employment opportunities are extensive. The work of electrical engineers can be seen in the homes we live in, in the automobiles we drive, in the computers we use, in numerically-controlled machines used by manufacturing companies, and in the security systems the federal government uses to ensure our national security.

Two outstanding sources of information about electrical engineering careers are the Sloan Career Cornerstone Center website at: www.careercornerstone.org/eleceng/eleceng.htm and the Institute of Electrical and Electronic Engineers (IEEE) website at: www.ieee.org

As you will see at the IEEE website, the IEEE is organized into the following 38 technical societies:

Technical Societies of the IEEE

Aerospace and electronic systems	Antennas and propagation
Broadcast technology	Circuits and systems
Communications	Electromagnetic compatibility
Ultrasonics, ferroelectrics, and frequency control	Components, packaging, and manufacturing technology
Computer	Control systems
Consumer electronics	Education
Computational intelligence	Intelligent transportation systems
Dielectrics and electrical insulation	Electron devices
Engineering in medicine and biology	Industrial electronics
Geoscience and remote sensing	Information theory
Industry applications	Photonics
Instrumentation and measurement	Microwave theory and techniques
Magnetics	Oceanic engineering

Nuclear and plasma sciences	Power & Energy
Power electronics	Solid state circuits
Professional communications	Reliability
Robotics and automation	Signal processing
Social implications of technology	Systems, man, and cybernetics
Product safety engineering	Vehicular technology

The listing of IEEE societies should give you an idea of the scope encompassed by the electrical engineering field. Within electrical engineering programs of study, the above 38 technical areas are generally organized under six primary specialties:

Computer Engineering*
Electronics
Communications
Power
Controls
Instrumentation

[*As previously noted, computer engineering will be discussed later in this section as a separate engineering discipline.]

Electronics deals with the design of circuits and electric devices to produce, process, and detect electrical signals. Electronics is rapidly changing and becoming increasingly important because of new advances in microelectronics. Most notable among these has been the doubling of the number of components that can be placed on a given surface area every two years for the past four decades, which has led to much smaller and more powerful devices. Our standard of living has significantly improved due to the advent of semiconductors and integrated circuits (ICs). Semiconductor products include not just digital ICs but also analog chips, mixed-signal (analog and digital integrated) circuits, and radio-frequency (RF) integrated circuits.

Communications involves a broad spectrum of applications from consumer entertainment to military radar. Recent advances in personal communication systems (e.g., cellular telephones, personal assistants, and GPS systems) and video-conferencing, along with technological advances in lasers and fiber optics, are bringing about a revolution in the communications field, opening up possibilities that were not even dreamed of a few years ago: e.g., online video conferencing, international broadcasting of conferences and tutorials, real-time transfer of huge data files, and transmission of integrated voice/data/video files. Wireless communication allows people to communicate anywhere with anyone by voice, e-mail, text, or instant messaging; to send or receive pictures and data; and to access the Internet.

Power involves the generation, transmission, and distribution of electric power. Power engineers are involved with conventional generation systems such as hydroelectric, steam, and nuclear, as well as alternative generation systems such as solar, wind, ocean tides, and fuel cells. Power engineers are employed wherever electrical energy is used to manufacture or produce a product – petrochemicals, pulp, paper, textiles, metals, and rubber, for example. As such, power engineers must have in-depth knowledge about transmission lines, electric motors, and generators.

Controls engineers design systems that control automated operations and processes. Control systems generally compare a measured quantity to a desired standard and make whatever adjustments are needed to bring the measured quantity as close as possible to the desired standard. Control systems are used in regulating the temperature of our buildings, reducing the emissions from our cars and trucks, ensuring the quality of chemical and industrial processes, maintaining reliable electrical output from our power plants, and ensuring highly efficient and fault tolerant voice and data networks. Unmanned aerial vehicles (UAVs) represent a major new and challenging application of controls engineering.

Instrumentation involves the use of electronic devices, particularly transducers, to measure such parameters as pressure, temperature, flow rate, speed, acceleration, voltage, and current. Instrumentation engineers not only conduct such measurements themselves; they also take part in processing, storing, and transmitting the data they collect.

E.2 MECHANICAL ENGINEERING

Mechanical engineering is currently the second largest engineering discipline (behind the combined discipline of electrical and computer engineering) in terms of the number of graduates annually and the third largest in terms of the number of employed engineers. According to the U.S. Department of Labor [1], of the 1.5 million engineers in 2011, 238,260 (16.1 percent) were mechanical engineers. Mechanical engineering is also one of the oldest and broadest engineering disciplines.

Mechanical engineers design tools, engines, machines, and other mechanical equipment. They design and develop power-producing machines such as internal combustion engines, steam and gas turbines, and jet and rocket engines. They also design and develop power-using machines such as refrigeration and air-conditioning equipment, robots, machine tools, materials handling systems, and industrial production equipment.

The work of mechanical engineers varies by industry and function. Specialties include, among others, applied mechanics, design, energy systems, pressure vessels and piping, heating, refrigeration, and air-conditioning systems, and manufacturing.

Mechanical engineers also design tools needed by other engineers for their work.

The American Society of Mechanical Engineers (ASME) lists 35 technical divisions.

Technical Divisions of the ASME

Applied mechanics	Internal combustion engines
Bioengineering	Power
Fluids engineering	Nuclear engineering
Heat transfer	Microelectromechanical systems (MEMS)
Tribology	Advanced energy systems
Aerospace	Fluid power systems and technology
Environmental engineering	Noise control and acoustics
Pipeline systems	Computers and information in engineering
Nondestructive evaluation	Information storage & processing systems
Management	Materials handling engineering
Solar energy	Safety engineering and risk analysis
Process industries	Plant engineering and maintenance
Materials	Technology and society
Petroleum	Materials and energy recovery
Manufacturing engineering	Ocean, offshore, and arctic engineering
Design engineering	Pressure vessels and piping
Rail transportation	Electronic and photonic packaging
	Dynamic systems and control

I'm sure this is a daunting list, but it is only the "tip of the iceberg." Each of these technical divisions is divided into a number of technical committees. For example, the Applied Mechanics Division is organized into 18 technical committees, each representing a subspecialty within the mechanical engineering field. A list of these committees will give you an idea about the applied mechanics area:

Composite Materials
Computing in Applied Mechanics
Dynamics and Control of Structures and Systems
Dynamic Response and Failure of Materials
Fracture and Failure Mechanics
Elasticity
Experimental Mechanics
Geomechanics
Materials Processing and Manufacturing

Fluid Mechanics
Instability in Solids & Structures
Transportation
Applied Mechanics Education
Uncertainty, Probabilistics and Reliability
Integrated Structures
Mechanics in Biology and Medicine
Fluid-Structure Interaction
Soft Materials

Within mechanical engineering study, these numerous technical fields and subspecialties are generally grouped into three broad areas:

Energy

Structures and motion in mechanical systems

Manufacturing

Energy involves the production and transfer of energy, as well as the conversion of energy from one form to another. Mechanical engineers in this area design and operate power plants, study the economical combustion of fuels, design processes to convert heat energy into mechanical energy, and create ways to put that mechanical energy to work. Mechanical engineers in energy-related fields also design heating, ventilation, and air-conditioning systems for our homes, offices, commercial buildings, and industrial plants. Some develop equipment and systems for the refrigeration of food and the operation of cold storage facilities; others design "heat exchange" processes and systems to transfer heat from one object to another. Still others specialize in production of energy from alternative sources such as solar, geothermal, and wind.

The second major area of mechanical engineering study involves the **design of structures** and the **motion of mechanical systems**. Mechanical engineers in these areas contribute to the design of automobiles, trucks, tractors, trains, airplanes, and even interplanetary space vehicles. They design lathes, milling machines, grinders, and drill presses used in the manufacture of goods. They help design the copying machines, faxes, personal computers, and related products that have become staples in our business and home offices. They are involved in the design of the many medical devices, systems, and equipment that help keep us healthy (and, in some cases, alive). Indeed, every piece of machinery that touches our lives, directly or indirectly, has been designed by a mechanical engineer.

Manufacturing, the third area of mechanical engineering study, is the process of converting raw materials into a final product. To take this process from start to finish, a variety of equipment, machinery, and tools is needed. Designing and building equipment and machines are what the manufacturing area of mechanical engineering entails. Put simply, mechanical engineers in this area design and manufacture the machines that make machines. They also design manufacturing processes, including automation and robotics, to help make the production of manufactured goods as efficient, cost-effective, and reliable as possible.

If you are interested in learning more about mechanical engineering, check out the ASME website: www.asme.org.

E.3 CIVIL ENGINEERING

Civil engineering is the currently the third largest engineering field in terms of yearly graduates and the second largest in terms of working engineers. According to the U.S. Department of Labor [1], of the 1.5 million engineers working in the U.S. in 2011, 254,130 (17.2 percent) were civil engineers. Civil engineering is the oldest branch of engineering, with major civil engineering projects dating back more than 5,000 years.

Today, civil engineers plan, design, and supervise the construction of facilities essential to modern life. Projects range from high-rise buildings to mass transit systems, from airports to water treatment plants, from space telescopes to off-shore drilling platforms.

The American Society of Civil Engineers (ASCE) is organized into 10 technical divisions and councils that develop and disseminate technical information through conferences, book and journals, and policies and standards:

Technical Divisions of ASCE

Aerospace	Cold regions
Computing and information technology	Energy
Lifeline earthquake engineering	Forensic engineering
Disaster risk management	Wind engineering
Geomatics	Pipelines

Within civil engineering study, these 10 technical areas are generally organized into seven academic specialties:

Structural engineering

Transportation engineering

Environmental engineering

Water resources engineering

Geotechnical engineering

Surveying

Construction engineering

Structural engineers design all types of structures: bridges, buildings, dams, tunnels, tanks, power plants, transmission line towers, offshore drilling platforms, and space satellites. Their primary responsibility is to analyze the forces that a structure would encounter and develop a design to withstand those forces. A critical part of this design process involves the selection of structural components, systems, and materials that would provide adequate strength, stability, and durability. Structural dynamics is a specialty within structural engineering that accounts for dynamic forces on structures, such as those resulting from earthquakes.

Transportation engineers are concerned with the safe and efficient movement of both people and goods. They thus play key roles in the design of highways and streets, harbors and ports, mass transit systems, airports, and railroads. They are also involved in the design of systems to transport goods such as gas, oil, and other commodities.

Environmental engineers are responsible for monitoring, controlling, preventing, and eliminating air, water, and land pollution. To these ends, they are typically involved in the design and operation of water distribution systems, waste water treatment facilities, sewage treatment plants, garbage disposal systems, air quality control programs, recycling and reclamation projects, toxic waste cleanup projects, and pesticide control programs.

Water resources is, by its very title, an engineering specialty focused on water-related problems and issues. The work of engineers in this area includes the operation of water availability and delivery systems, the evaluation of potential new water sources, harbor and river development, flood control, irrigation and drainage projects, coastal protection, and the construction and maintenance of hydroelectric power facilities.

Geotechnical engineers analyze the properties of soil and rocks over which structures and facilities are built. From the information their analyses yield, geotechnical engineers are able to predict how the ground material would support or otherwise affect the structural integrity of a planned facility. Their work is thus vital to the design and construction of earth structures (dams and levees), foundations of buildings, offshore platforms, tunnels, and dams. Geotechnical engineers also evaluate the settlement of buildings, stability of slopes and fills, seepage of groundwater, and effects of earthquakes.

Engineers involved in **Surveying** are responsible for "mapping out" construction sites and their surrounding areas before construction can begin. They locate property lines and determine right-of-ways, while also establishing the alignment and proper placement of the buildings to be constructed. Current surveying practice makes use of modern technology, including satellites, aerial and terrestrial photogrammetry, and computer processing of photographic data.

Construction engineers use both technical and management skills to plan and build facilities – such as buildings, bridges, tunnels, and dams – that other engineers and architects designed. They are generally responsible for such projects from start to finish: estimating construction costs, determining equipment and personnel needs, supervising the construction, and, once completed, operating the facility until the client assumes responsibility. Given the breadth of such projects, construction engineers must be knowledgeable about construction methods and equipment, as well as principles of planning, organizing, financing, managing, and operating construction enterprises.

You can find lots of useful information about civil engineering at the ASCE website at: www.asce.org.

E.4 COMPUTER ENGINEERING

Compared to the three previous engineering disciplines we have discussed – electrical engineering, mechanical engineering, and civil engineering – computer engineering a relatively new field. The first accredited computer engineering program in the U.S. was established in 1971 at Case Western Reserve University. Since then, computer engineering has experienced rapid growth. It currently ranks fourth in terms of B.S. degrees conferred among engineering disciplines (see table on Pages 60-61). That growth is expected to continue in response to the needs of a world that will become increasingly "computer centered."

According to the U.S. Department of Labor statistics [1], show that in 2011 71,990 (4.9 percent) of the 1.5 million engineers in the country were computer engineers. One indication of the current demand for computer engineers is that the average starting salary for computer engineering graduates in 2010/11 was $68,500, compared to the average for all engineering graduates of $60,638 [2].

Computer engineering, which had its beginnings as a specialty or option within electrical engineering, and continues to rely on much of the same basic knowledge that the EE curriculum teaches, developed into a discipline of its own because of the growing need for specialized training in computer technology. To respond to this need, computer specialists in electrical engineering had to step up their research and course development, which increasingly brought them into contact with computer scientists. Today, although computer engineering and computer science remain separate disciplines, the work of computer engineers and computer scientists is often inseparable – or, more accurately, interdependent. One writer from IEEE aptly explains the relationship between computer engineering and computer science in terms of a "continuum":

> *At one pole is computer science, primarily concerned with theory, design, and implementation of software. It is a true engineering discipline, even though the product is an intangible – a computer program. At the other pole is computer engineering, primarily concerned with firmware (the microcode that controls processors) and hardware (the processors themselves, as well as entire computers). It is not possible, however, to draw a clear line between the two disciplines; many practitioners function at least to some extent as both computer engineers and computer scientists.*

While explaining the overlapping nature of the work of computer engineers and scientists, the passage also points out the major difference between them. That is, computer engineers focus more on computer hardware; computer scientists focus more on computer software.

I assume that most of you are already somewhat familiar with these terms. Given their importance in this discussion, however, we'll digress briefly to clarify them. "Hardware" refers to the machine itself: the chips, circuit boards, networks, devices, and other physical components of a computer. "Software" refers to the programs that tell the computer what to do and how to do it. A software program is literally a set of instructions, rules, parameters, and other guidelines, encoded in a special "language" that the hardware can read and then execute. A computer therefore needs both hardware and software, developed in tandem, in order to perform a given function.

As hardware specialists, computer engineers are concerned with the design, construction, assessment, and operation of high-tech devices ranging from tiny microelectronic integrated-circuit chips to powerful systems that utilize those chips and efficient telecommunication systems that interconnect those systems.

Applications include consumer electronics (CD and DVD players, televisions, stereos, microwaves, gaming devices) and advanced microprocessors, peripheral equipment (magnetic disks and tapes, optical disks, RAM, ROM, disk arrays, printers and plotters, visual displays, speech and sound software, modems, readers and scanners, keyboards, mouse devices, and speech input systems), systems for portable, desktop and client/server computing, and communications devices (cellular phones, pagers, personal digital assistants).

Other applications include distributed computing environments (local and wide area networks, wireless networks, internets, intranets), and embedded computer systems (such as aircraft, spacecraft, and automobile control systems in which computers are embedded to perform various functions). A wide array of complex technological systems, such as power generation and distribution systems and modern processing and manufacturing plans, rely on computer systems developed and designed by computer engineers.

As noted above, however, the work of computer engineers and computer scientists typically involves much crossover. That is, for any given design project, computer engineers' ability to deliver the appropriate hardware depends on their understanding of computer scientists' software requirements. As a result, they often participate in the development of the software – and may even create software of their own to support computer scientists' programs. Similarly, computer scientists' ability to deliver a viable software program depends on their knowledge of hardware. They thus play a critical role in facilitating computer engineers' design and development of the

necessary physical components, systems, and peripheral devices.

An excellent study, "Computer Engineering 2004: Curriculum Guidelines for Undergraduate Degree Programs in Computer Engineering," carried out by a joint task force of the IEEE Computer Society and the Association for Computing Machinery, lists 18 primary disciplines that make up the body of knowledge for computer engineering [3].

These are:

> *Algorithms*
> *Computer architecture and organization*
> *Computer systems engineering*
> *Circuits and signals*
> *Database systems*
> *Digital logic*
> *Digital signal processing*
> *Electronics*
> *Embedded systems*
> *Human-computer interaction*
> *Computer networks*
> *Operating systems*
> *Programming fundamentals*
> *Social and professional issues*
> *Software engineering*
> *VLSI design and fabrication*
> *Discrete structures*
> *Probability and statistics*

Each of these knowledge areas is further subdivided into more specific topics. To give a sense of the scope of the field of computer engineering, for example, the area of "Electronics" includes the following topics:

> *Electronic properties of materials*
> *Diodes and diode circuits*
> *MOS transistors and biasing*
> *MOS logic families*
> *Bipolar transistors and logic families*
> *Design parameters and issues*
> *Storage elements*
> *Interfacing logic families and standard buses*
> *Operational amplifiers*
> *Circuit modeling and simulation*
> *Data conversion circuits*
> *Electronic voltage and current sources*
> *Amplifier design*
> *Integrated circuit building blocks*

Because computer engineers work closely with systems analysts and computer scientists and the three disciplines overlap significantly, the fields of systems analyst and computer scientist are discussed in the following two sections.

SYSTEMS ANALYST. Whatever work computer engineers do, it is typically generated by some company or government need, which if you recall the engineering design process, leads to a problem definition and specifications. Identifying these needs and initiating design projects to solve them is the responsibility of **systems analysts.** Comprising another fast growing field of computer technology, systems analysts are charged with planning, developing, and selecting new computer systems, or modifying existing programs to meet the needs of an organization. Although their training is in management information systems (as opposed to engineering or computer science), systems analysts are highly computer-literate specialists who work as corporate "watchdogs" to ensure that their company is realizing the maximum benefits from its investment in equipment, personnel, software, and business practices.

COMPUTER SCIENTISTS. Finally, since computer engineers work so frequently – and so closely – with computer scientists, a brief overview of that field would provide a fitting conclusion to our discussion of computer engineering. Computer scientists have already been distinguished as the software experts in the general field of computer technology. As software specialists, their work tends to be highly theoretical, involving extensive, complex applications of math and science principles, algorithms, and 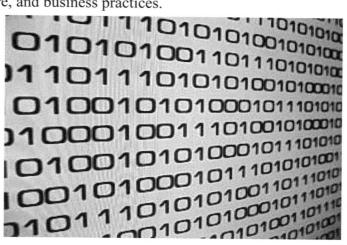 other computational processes. However, we have also seen that their theoretical work requires an understanding of the many physical components, processes, and functional requirements of computers. The Computing Sciences Accreditation Board (CSAB) begins its definition of the discipline as one that

> involves the understanding and design of computers and computational processes . . . The discipline ranges from theoretical studies of algorithms to practical problems of implementation in terms of computational hardware and software.

The definition continues, but this brief passage is sufficient to describe the computer science discipline.

Computer science programs in the U.S. are accredited by the Accreditation Board for Engineering and Technology (ABET) in conjunction with the Computer Science Accreditation Board (CSAB), an organization representing the three largest computer and computer-related technical societies: the Association of Computing Machinery (ACM), the Association for Information Systems (AIS), and the IEEE-Computer Society. During 2011/12, 9,700 computer science degrees were awarded by the 258 universities having accredited computer science programs.

Computer science is a rapidly expanding field. As a result, computer science curricula is in a constant state of flux. To get an idea of the current and future state of computer science education, the Joint Task for on Computing Curricula of the Association for Computing Machinery (ACM) and the IEEE Computer Society published "Computer Science Curriculum 2013" [4], which put forth 18 knowledge areas that should be included in all computer science programs:

Algorithms and Complexity
Architecture and Organization
Computational Science
Discrete Structures
Graphics and Visual Computing
Human-Computer Interaction
Information Assurance and Security
Information Management

Intelligent Systems
Networking and Communications
Operating Systems
Platform-based Development
Parallel and Distributed Computing
Programming Languages
Software Development Fundamentals
Software Engineering
Systems Fundamentals
Social and Professional Issues

This list is comprised of a broad-based core and advanced-level subjects.

The core provides basic coverage of algorithms, data structures, software design, concepts of programming languages, and computer organization and architecture.

Examples of advanced areas include algorithms and data structures, artificial intelligence and robotics, computer networks, computer organization and architecture, database and information retrieval, human-computer interaction, numerical and symbolic computation, operating systems, programming languages, software methodology and engineering, and theory of computation. (Note that many of these areas are the same ones listed above for computer engineering. These shared areas only reinforce the overlap and similarities between computer science and computer engineering.)

More information about computing and computer science can be found at the following websites:

Computer Science Accreditation Board	www.csab.org
IEEE Computer Society	www.computer.org
Association of Computing Machines	www.acm.org
Association for Information Systems	www.aisnet.org

E.5 CHEMICAL ENGINEERING

Chemical engineers combine their engineering training with a knowledge of chemistry to transform the laboratory work of chemists into commercial realities. They are most frequently involved in designing and operating chemical production facilities and manufacturing facilities that use chemicals (or chemical processes) in their production of goods.

The scope of the work of chemical engineers is reflected by the industries that employ them – manufacturing, pharmaceuticals, health-care, design and construction, pulp and paper, petrochemicals, food processing, specialty chemicals, micro-electronics, electronic and advanced materials, polymers, business services, biotechnology, and environmental health and safety.

The work of chemical engineers can be seen in a wide variety of products that affect our daily lives, including plastics, building materials, food products, pharmaceuticals, synthetic rubber, synthetic fibers, and petroleum products (e.g., shampoos, soaps, cosmetics, shower curtains, and molded bathtubs).

Chemical engineers also play a major role in keeping our environment clean by creating ways to clean up the problems of the past, prevent pollution in the future, and extend our shrinking natural resources. Many play equally important roles in helping to eliminate world hunger by developing processes to produce fertilizers economically.

The scope of chemical engineering is reflected by the 20 divisions and forums of the American Institute of Chemical Engineers (AIChE):

Divisions and Forums of the AIChE

Catalysis and Reaction Engineering	Environmental
Computational Molecular Science & Engineering	Food, Pharmaceutical & Bioengineering
Computing & Systems Technology	Fuels & Petrochemicals
Upstream Engineering and Flow Assurance	Management
Materials Engineering & Sciences	Nuclear Engineering
Nanoscale Science Engineering	Particle Technology
North American Mixing	Process Development
Safety & Health	Separations
Sustainable Engineering	Forest Bioproducts
Chemical Engineering and the Law	Education

You can learn more about chemical engineering by visiting the AIChE webpage at: www.aiche.org.

An excellent source of practical information for students considering chemical engineering can be found by googling "Answers to Career FAQs by Chemical Engineer Members."

E.6 BIOENGINEERING/BIOMEDICAL ENGINEERING

Bioengineering is a wide-ranging field, alternately referred to as biomedical engineering, which was created some 40 years ago by the merging interests of engineering and the biological/medical sciences. It is perhaps the most rapidly growing field of engineering as indicated by the fact that the number of accredited biomedical engineering programs in the U.S. grew from 36 in 2005 to 73 in 2012. Of the 1.5 million engineers working in the U.S. in 2011, 16,590 identified themselves as biomedical engineers.

Bioengineers work closely with health professionals in designing diagnostic and therapeutic devices for clinical use, designing prosthetic devices, and developing biologically compatible materials. Pacemakers, blood analyzers, cochlear implants, medical imaging, laser surgery, prosthetic implants, and life support systems are just a few of the many products and processes that have resulted from the team efforts of bioengineers and health professionals.

Biomedical engineering is a field in continual change, creating new areas due to rapid advancement in technology. However, some of the well established specialty areas within the field are bioinstrumentation; biomaterials; biomechanics; cellular, tissue and genetic engineering; clinical engineering; medical imaging; orthopedic surgery; rehabilitation engineering; and systems physiology. For more information about careers in biomedical engineering, see the Biomedical Engineering Society (BMES) webpage at: www.bmes.org.

For some excellent information about the Biomedical Engineering field, click on "About BMES" and then click on "FAQs About Biomedical Engineering."

E.7 INDUSTRIAL ENGINEERING

Industrial engineers determine the most effective ways for an organization to use its various resources – people, machines, materials, information, and energy – to develop a process, make a product, or provide a service. Their work does not stop there, however, for they also design and manage the quality control programs that monitor the production process at every step. They also may be involved in facilities and plant design, along with plant management and production engineering.

These multiple responsibilities of an industrial engineer require knowledge not only of engineering fundamentals, but also of computer technology and management practices. At first glance, the industrial engineer might be seen as the engineering equivalent of a systems analyst – except that the industrial engineer plays many more roles and has a much wider window of career opportunities.

Perhaps the single most distinguishing characteristic of industrial engineers is their involvement with the human and organizational aspects of systems design. Indeed, the Institute of Industrial Engineers (IIE) describes industrial engineering as "The People-Oriented Engineering Profession."

Sixty percent of industrial engineers are employed by manufacturing companies, but industrial engineers can be found in every kind of institution (e.g., financial, medical, agricultural, governmental) and commercial field (e.g., wholesale and retail trade, transportation, construction, entertainment, etc.).

Given its breadth of functions in so many areas, industrial engineering has been particularly impacted by recent advances in computer technology, automation of manufacturing systems, developments in artificial intelligence and database systems, management practices (as reflected by the "quality movement"), and the increased emphasis on strategic planning.

The scope of industrial engineering can be gleaned from the 11 technical societies and divisions of the IIE:

Technical Societies and Divisions of the IIE

Engineering & Management Systems	Quality Control & Reliability Engineering
Work Systems	Applied Ergonomics
Computer & Information Systems	Construction

Engineering Economy	Lean
Operations Research	Process Industries
Health Systems	

The unique role of industrial engineering in improving productivity, reducing costs, enhancing customer satisfaction, and achieving superior quality is perhaps reflected best by the fact that the IIE includes a Lean Division. To learn more about industrial engineering, visit the IIE website at: www.iienet2.org.

E.8 AEROSPACE ENGINEERING

Aerospace engineers design, develop, test, and help manufacture commercial and military aircraft, missiles, and spacecraft. They also may develop new technologies in commercial aviation, defense systems, and space exploration. In this work, they tend to focus on one type of aerospace product such as commercial transports, helicopters, spacecraft, or rockets. Specialties within aerospace engineering include aerodynamics, propulsion, thermodynamics, structures, celestial mechanics, acoustics, and guidance and control systems.

For more information about aerospace engineering, go to the American Institute of Aeronautics and Astronautics (AIAA) webpage at: www.aiaa.org.

E.9 OVERVIEW OF OTHER ENGINEERING DISCIPLINES

The following sections provide an overview of each of the more specialized, non-traditional engineering disciplines.

MATERIALS ENGINEERING/METALLURGICAL ENGINEERING. Materials engineers are generally responsible for improving the strength, corrosion resistance, fatigue resistance, and other

characteristics of frequently used materials. The field encompasses the spectrum of materials: metals, ceramics, polymers (plastics), semiconductors, and combinations of materials called composites. Materials engineers are involved in selecting materials with desirable mechanical, electrical, magnetic, chemical, and heat transfer properties that meet special performance requirements. Examples are graphite golf club shafts that are light but stiff, ceramic tiles on the space shuttle that protect it from burning up during reentry into the atmosphere, and the alloy turbine blades in a jet engine.

The materials field offers unlimited possibilities for innovation and adaptation through the ability to actually engineer, or create, materials to meet specific needs. This engineering can be carried out at the atomic level through the millions of possible combinations of elements. It can also be done on a larger scale to take advantage of unique composite properties that result from microscopic-scale combinations of metals, ceramics and polymers, such as in fiber reinforcement to make a graphite fishing rod or, on a slightly larger scale, for steel-belted radial tires. Finally, it can be practiced on an even larger scale with bridges, buildings, and appliances.

Metallurgical engineers deal specifically with metal in one of the three main branches of metallurgy: extractive, physical, and mechanical. Extractive metallurgists are concerned with removing metals from ores, and refining and alloying them to obtain useful metal. Physical metallurgists study the nature, structure, and physical properties of metals and their alloys, and design methods for processing them into final products. Mechanical metallurgists develop and improve metal-working processes such as casting, forging, rolling, and drawing.

For more information about materials engineering, go to the Materials Information Society website at: www.asminternational.org.

ARCHITECTURAL ENGINEERING. Architectural engineers work closely with architects on the design of buildings. Whereas the architect focuses primarily on space utilization and aesthetics, the architectural engineer is concerned with safety, cost, and sound construction methods.

Architectural engineers focus on areas such as the following: the structural integrity of buildings to anticipate earthquakes, vibrations and wind loads; the design and analysis of heating, ventilating and air conditioning systems; efficiency and design of plumbing, fire protection and electrical systems; acoustic and lighting planning; and energy conservation issues.

For more information on architectural engineering, go to the ASCE Architectural Engineering Institute website at: www.asce.org/aei.

AGRICULTURAL ENGINEERING. Agricultural engineers are involved in every aspect of food production, processing, marketing, and distribution. Agricultural engineers design and develop agricultural equipment, food processing equipment, and farm structures. Major technical areas of agricultural engineering include food processing; information and electrical technologies; power and machinery, structures, soil and water; forestry; bioengineering; and aqua culture. With their technological knowledge and innovations, agricultural engineers have literally revolutionized the farming industry, enabling farmers today to produce approximately ten times more than what they could just 100 years ago.

To learn more about agricultural engineering, visit the American Society of Agricultural and Biological Engineers (ASABE) website at: www.asabe.org.

SYSTEMS ENGINEERING. Systems engineers are involved with the overall design, development, and operation of large, complex systems. Their focus is not so much on the individual components that comprise such systems; rather, they are responsible for the integration of each component into a complete, functioning "whole." Predicting and overseeing the behavior of large-scale systems often involves knowledge of advanced mathematical and computer-based techniques, such as linear programming, queuing theory, and simulation.

Find out more about systems engineering by visiting the International Council on Systems Engineering (INCOSE) website at: www.incose.org. Click on "Education & Careers." Then click on "Careers in SE."

ENVIRONMENTAL ENGINEERING.   Environmental engineers, relying heavily on the principles of biology and chemistry, develop solutions to environmental problems. Environmental engineers work in all aspects of environmental protection including air pollution control, industrial hygiene, radiation protection, hazardous waste management, toxic materials control, water supply, wastewater management, storm water management, solid waste disposal, public health, and land management.

Environmental engineers conduct hazardous-waste management studies in which they evaluate the significance of the hazard, offer analysis on treatment and containment, and develop regulations to prevent mishaps. They design municipal water supply and industrial wastewater treatment systems. They conduct research on proposed environmental projects, analyze scientific data, and perform quality control checks. They provide legal and financial consulting on matters related to the environment. For information about environmental engineering, see the American Academy of Environmental Engineers (AAEE) website at: www.aaee.net/Website/Careers.htm.

MARINE ENGINEERING/OCEAN ENGINEERING/NAVAL ARCHITECTURE. Naval architects, marine engineers, and ocean engineers design, build, operate, and maintain ships and other waterborne vehicles and ocean structures as diverse as aircraft carriers, submarines, sailboats, tankers, tugboats, yachts, underwater robots, and oil rigs. These interrelated professions address our use of the seas and involve a variety of engineering and physical science skills, spanning disciplines that include hydrodynamics, material science, and mechanical, civil, electrical, and ocean engineering.

Marine engineers are responsible for selecting ships' machinery, which may include diesel engines, steam turbines, gas turbines, or nuclear reactors, and for designing mechanical, electrical, fluid, and control systems throughout the vessel. Some marine engineers serve aboard ships to operate and maintain these systems. Ocean engineers study the ocean environment to determine its effects on ships and other marine vehicles and structures. Ocean engineers may design and operate stationary ocean platforms, or manned or remote-operated sub-surface vehicles used for deep sea exploration. Naval architects are involved with basic ship design, starting with hull forms and overall arrangements, power requirements, structure, and stability. Some naval architects work in shipyards, supervising ship construction, conversion, and maintenance.

To learn more about marine engineering, ocean engineering, and naval architecture, visit the Society of Naval Architects & Marine Engineers website at: www.sname.org.

PETROLEUM ENGINEERING. Petroleum engineers work in all capacities related to petroleum (gas and oil) and its byproducts. These include designing processes, equipment, and systems for locating new sources of oil and gas; sustaining the flow of extant sources; removing, transporting, and storing oil and gas; and refining them into useful products.

For more information about petroleum engineering, go to the Society of Petroleum Engineering (SPE) website at: www.spe.org.

NUCLEAR ENGINEERING. Nuclear engineers are concerned with the safe release, control, utilization, and environmental impact of energy from nuclear fission and fusion sources. Nuclear engineers are involved in the design, construction, and operation of nuclear power plants for power generation, propulsion of nuclear submarines, and space power systems. Nuclear engineers are also involved in processes for handling nuclear fuels, safely disposing radioactive wastes, and using radioactive isotopes for medical purposes.

For information about nuclear engineering and nuclear technology visit the American Nuclear Society (ANS) website at: www.ans.org.

MINING ENGINEERING/GEOLOGICAL ENGINEERING. The work of mining and geological engineers is similar to that of petroleum engineers. The main difference is the target of their efforts. That is, mining and geological engineers are involved in all aspects of discovering, removing, and processing minerals from the earth. The mining engineer designs the mine layout, supervises its construction, and devises systems to transport minerals to processing plants. The mining engineer also devises plans to return the area to its natural state after extracting the minerals.

For more information on mining engineering, visit the Society of Mining, Metallurgy & Exploration website at: www.smenet.org.

MANUFACTURING ENGINEERING. Manufacturing engineers are involved in all aspects of manufacturing a product. These include studying the behavior and properties of required materials, designing appropriate systems and equipment, and managing the overall manufacturing process.

Find out more about manufacturing engineering at the Society of Manufacturing Engineers (SME) website at: www.sme.org.

CERAMIC ENGINEERING. Ceramic engineers direct processes that convert nonmetallic minerals, clay, or silicates into ceramic products. Ceramic engineers work on products as diverse as glassware, semiconductors, automobile and aircraft engine components, fiber-optic phone lines, tiles on space shuttles, solar panels, and electric power line insulators.

You can learn more about ceramic engineering at the American Ceramic Society website: www.ceramics.org.

SOFTWARE ENGINEERING. Software engineering involves the creation of software using a process similar to other engineering disciplines. Programming is only one phase (construction) of software engineering. There are many other aspects of the software engineering process, such as requirements definition, architectural design, and quality assurance, which need to be applied in order to develop reliable software on time and within budget constraints.

For a detailed description of software engineering, visit the Wikipedia website: www.en.wikipedia.org/wiki/Software_engineering. To learn more about software engineering, visit the IEEE Computer Society website at: www.computer.org.

CONSTRUCTION ENGINEERING.

Construction engineering is a professional discipline that deals with the designing, planning, construction, and management of infrastructures such as highways, bridges, airports, railroads, buildings, dams, and utilities. Construction Engineers are unique such that they are a cross between civil engineers and construction managers. Construction engineers learn the designing aspect much like civil engineers and construction site management functions much like construction managers.

For a more detailed description of construction engineering, go to the Wikipedia website at: www.en.wikipedia.org/wiki/Construction_engineering.

ENGINEERING MANAGEMENT.

Engineering management is concerned with the application of engineering principles to business practice. Engineering management is a career that brings together the technological problem-solving savvy of engineering and the organizational, administrative, and planning abilities of management in order to oversee complex enterprises from conception to completion. Potential career areas include product development, project management, materials management, facilities oversight, quality assurance, research, and manufacturing.

To learn more about engineering management, go to the American Society for Engineering Management website at: www.asem.org.

OPTICS ENGINEERING.

Optical engineers design components of optical instruments such as lenses, microscopes, telescopes, and other equipment that utilize the properties of light. Other devices include laser printers, fiber optic communication, internet switches, fiber optic telephone lines, compact disc players, credit cards bearing holograms, grocery checkout scanners, computers, and eye surgery. Opportunities for graduates in optical engineering are available in industries such as automated inspection, consumer electronics, fiber optic communications, optical instrumentation, laser devices, radar systems, and data storage.

SURVEYING AND GEOMATICS ENGINEERING.

Surveying and geomatics engineers manage the global spatial infrastructure. This effort includes real property boundary determination, digital mapping, Geographic Information Systems (GIS), Global Positioning Systems (GPS), remote sensing, photogrammetric mapping, applications programming, project management, and construction layout activities. Students are exposed to a wide selection of specialized equipment while acquiring a solid theoretical background.

For a more detailed description of geomatics engineering, see the Wikipedia website at: www.en.wikipedia.org/wiki/Geomatics_engineering.

TELECOMMUNICATIONS ENGINEERING. Telecommunications engineers design, develop, test, and debug software and hardware products for communications applications. These products range from modems and encoders to computer-assisted engineering programs for schematic cabling projects; modeling programs for cellular and satellite systems; and programs for telephone options, such as voice mail, email, and call waiting.

A source of information about telecommunications engineering is the IEEE Communications Society website at: www.comsoc.org.

References

1. "May 2011 National Occupational Employment and Wage Estimates - United States," U. S. Department of Labor, Bureau of Labor Statistics. www.bls.gov/oes/current/oes_nat.htm.

2. "NACE Salary Survey: Starting Salaries for New College Graduates," September, 2012, National Association of Colleges and Employers (www.naceweb.org).

3. "Computer Engineering 2004: Curriculum Guidelines for Undergraduate Degree Programs in Computer Engineering," IEEE Computer Society, 2004.

4. "Computer Science Curriculum 2013," Joint Task Force on Computing Curricula Association for Computing Machinery IEEE Computer Society, February, 2012 (ai.stanford.edu/users/sahami/CS2013//strawman-draft/cs2013-strawman.pdf).

INDEX